Mini
Missy
Matta
Marcella
Moko and Matta
Marlon
Midas
Mitsou
May
Lola
Maillol

SLEEPY CAT FARM

A GARDENER'S JOURNEY

SLEEPY CAT FARM

A GARDENER'S JOURNEY

Text by Caroline Seebohm

Photographs by Curtice Taylor

Introduction by Ken Druse

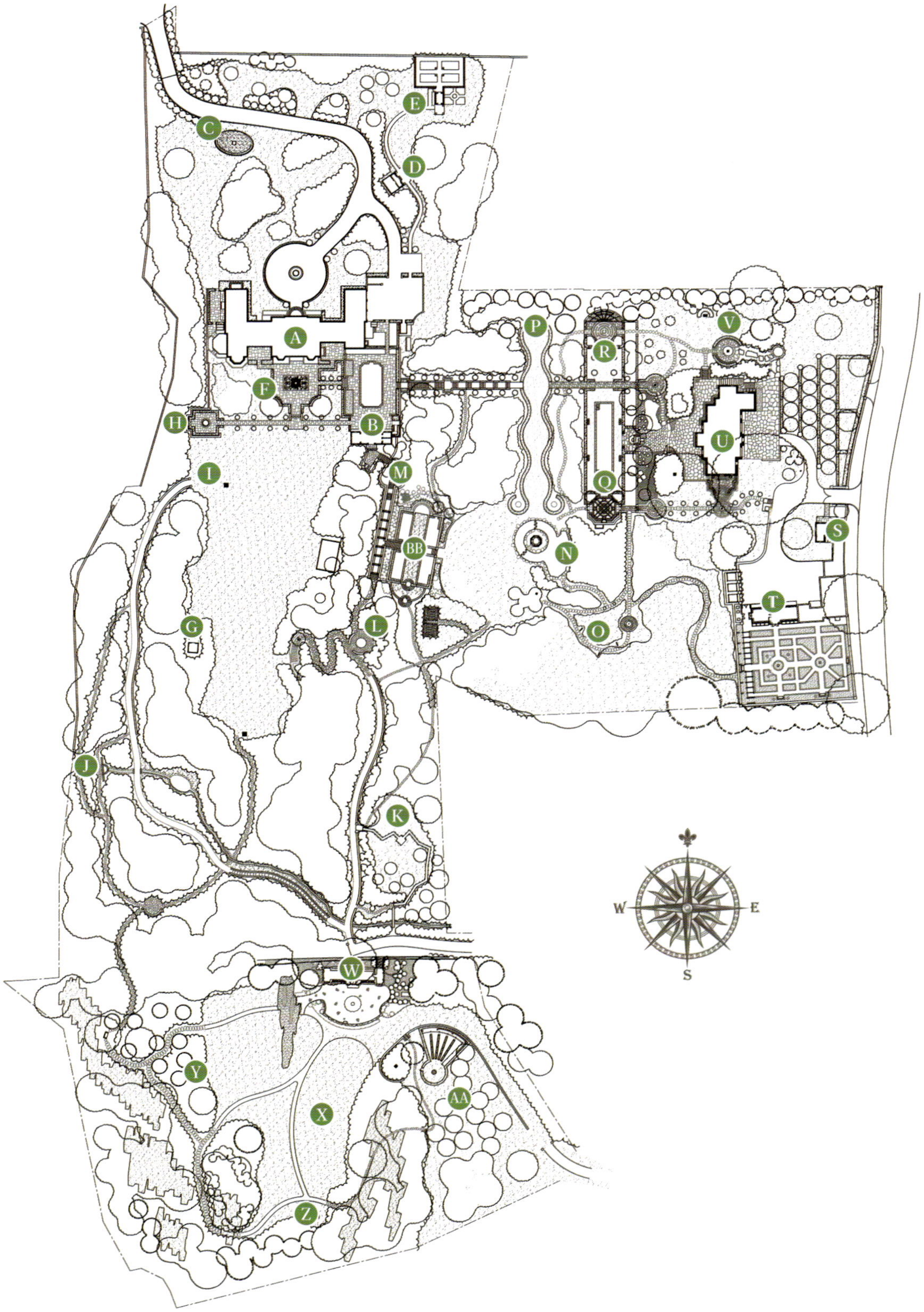

Garden Map

THE HOUSE

A. Main House
B. Pool House
C. Paris & Helen
D. Flower Garden
E. Upper Potager

THE LAWN & THE RIDGE

F. Parterre
G. Celestial Pavilion
H. Putti Fountain

THE GOLDEN PATH

I. The Path
J. The Grotto
K. Iris Garden & Spirit Walk
L. Narcissus Pool
M. Mercury's Arbor

EXPANDING HORIZONS

N. Chinese Pavilion
O. Prayer Wheels
P. Atlas
Q. Pebble Terrace
R. Cat Maze & Wisteria Arbor

CHANGES IN STYLE

S. Farm Office
T. Greenhouse & Vegetable Garden
U. The Barn
V. Cascade

NEW LAND, NEW IDEAS

W. Limonaia
X. Meadow & Lower Orchard
Y. Lower Woodland Walk
Z. Garden Seat
AA. Berry Wedge

THE NEVER-ENDING STORY

BB. Perennial Long Border Garden

CONTENTS

INTRODUCTION

Imagine that you had a magic wand, or a magic trowel, and your every garden dream could come true. It seems that everything Fred Landman has envisioned for Sleepy Cat Farm has come to pass. But it took more than a wave of a wand to make this place. It took time, talent, travel, and the tenacity of a guiding hand. And all of that on a challenging rocky ridge in the southwest corner of Connecticut.

I've been photographing and writing about gardens for more than thirty years. When I was asked to do a story on "a young garden" for a Connecticut magazine fifteen years ago, I have to admit I had reservations. I'd been sent to gardens before by editors who, all too often, hadn't seen the locations, and I ended up having to do a lot of tap dancing to get a story out of them.

It was around the same time that my colleague garden photographer Curtice Taylor ran into Fred Landman at Pearl's Oyster Bar in New York City. Although they hadn't seen each other since high school, Landman recognized him. Landman had heard Taylor was an accomplished photographer and suggested he visit Sleepy Cat Farm. Like me, Curtice Taylor had heard this kind of invitation before and had minimal expectations, but he had ways of determining a gardener's expertise.

"Planting any daffodils lately?" he asked. Landman answered in the affirmative. The next telling question was "Oh, about how many?"

Fred Landman confessed that it was a paltry purchase that year.

"Only 5,000," he answered.

View over the perennial border and lawn leading to the Porta Paradiso, the most recent addition to the Sleepy Cat gardens.

A date was set, and Curtice Taylor has been photographing the gardens of Sleepy Cat (named for the dozen resident felines) ever since. Many of his romantic photographs live on the pages of this book.

Like Taylor, I found a lot more than I expected. When I arrived, I passed by the side of the house along a paved walk leading to a gate with a tasteful colonial-style cannon ball and chain gravity closure. I began to get a little more excited and to let go of my writer's prejudice. But even this stylish hardware didn't alert me to what was to come. Through the gate, at the southeast corner of the house, was a swimming pool and across from that, a charming brick pool house.

When the renovated 1942 house was settling into its reincarnation, Landman felt that it, like a superb painting, needed the perfect frame. But he realized he could use some help and brought in Charles Stick, a Charlottesville, Virginia, landscape architect whose interest in historical garden design fit Fred Landman's aesthetic. The men have been collaborating ever since, now for more than twenty years.

Landman was my guide to the next area Stick designed—a formal parterre of boxwood behind the house—and on a secret spot with a marble sculpture of boys standing on a giant-clam shell. One was blowing a conch shell horn and the other wrestling a mythical dolphin. The group convened above a round pool encircled by trimmed boxwood. There was also a bench to contemplate the scene—the famous eponymous arts-and-crafts design from 1902 by architect Sir Edwin Lutyens—the Lutyens bench—found today in gardens from Sissinghurst in Kent, England, to Sleepy Cat Farm in Connecticut. This sculpture and bench rest below a gigantic Japanese snowbell tree, a *Styrax japonica* with arching branches smothered in white flowers.

There are benches throughout the garden, and Landman sits on them to contemplate the views. "I think benches are important," he said. It's essential that they look good where they are sited and that there is a good view from them as well.

Above: This marble sculptural group was discovered on the property and reconfigured as a fountain. A traditional Lutyens bench is placed in front of a yew hedge.

Above: A wisteria arbor shelters a Chippendale-style bench that overlooks the boxwood cat maze.

Above right: *L'Été sans bras*, a bronze figure by Aristide Maillol, is installed at the south end of the lawn.

One example, added recently, is the wide Chippendale-style bench beneath a massive semi-circular pergola enveloped by an equally impressive wisteria vine. There's a spectacular view from this spot beginning with a low hedge arrangement called the cat maze and continuing across a formal reflecting pool punctuated at each corner by columnar oaks and ending with Emile-Antoine Bourdelle's bronze sculpture *Guerrier allongé au glaive*, one of more than twenty such works of art at Sleepy Cat today.

Back when I first visited, I was ushered past one of Landman's earliest and best-known acquisitions, a Maillol bronze entitled *L'Été sans bras*. We next entered a shady area with a decomposed-granite path that wound through plantings alongside massive rock outcroppings. I could hear something: babbling water. (Landman has always liked the sound of moving water in a garden.) It was a Grotto designed as a classical ruin with gentle waterfall and stream trickling into a pool where a statue of Venus admired her reflection.

In the early days, the path led out of the woods. The narrowed space opened up to the sky. Lawn took over from the woodland understory, and there, in the clearing was an island in a pond filled with lotus, and on the island a Chinese-red painted pavilion. Landman told me the building was inspired by a trip to China and described it as a Ting, rather than a pagoda because it is a place for enjoying the garden, not a place of worship. Fifteen years ago, my tour ended there.

When Fred Landman traveled to places around the world, Charles Stick would recommend gardens to visit. Landman often developed new areas influenced by what he saw. He's gone to gardens in North America—for instance, Monticello in Virginia— and others throughout Europe—Russell Page's Giardini La Mortella in Ischia, Marie Antoinette's vegetable garden at Yvelines, France. The Grotto looks a lot like one at Rousham House in Oxfordshire. These architectural constructions are not exact copies as much as homages adapted for modern New England.

The places Landman has visited and loved all seem to have things in common: water for sure, but more, personality. The gardens were all constructed around what were once private homes. And like Sleepy Cat Farm, they were landscapes to wander and discover. Remarkable structures reveal themselves bit by bit along what is often referred to as the Golden Path. There are special plants, too, and impressive plantings such as the undulating waves of hornbeam hedge.

Sleepy Cat Farm is occasionally open to the public to raise money for organizations such as the Garden Conservancy, founded by Frank Cabot. I wrote an article on Cabot's magnificent private garden in Quebec, Les Quatre-Vents (Four Winds), for the 100th anniversary issue of *House Beautiful*—the magazine founded by Frank Lloyd Wright in 1897. So much of Les Quatre Vents was created to delight and even surprise guests. It just happened to be about a century after that magazine premiered that Fred Landman decided to become a fulltime garden maker.

Frank Cabot's legacy of garden-making and the influence of the Conservancy he established have been felt all over North America, including my own garden in New Jersey and Sleepy Cat Farm. Landman considers Les Quatre Vents one of his greatest driving forces.

Above: Fred Landman greets two legendary gardeners, John Trexler and Marco Polo Stufano.

Opposite: Landman collecting tickets and suggesting a route for visitors to tour the garden during a Garden Conservancy Open Day.

On my most recent visit to Sleepy Cat Farm, he told me about another place that impressed him: Heywood Gardens in Ireland designed by Lutyens. The house is gone, as are most of the plantings likely created by Gertrude Jekyll, but the bones of Lutyens's formal garden remain. What struck Landman was the look on the faces of people touring. They smile, as do those who come to see Sleepy Cat Farm.

Artists need an audience, and no work is complete until it is shared. Landman gets that thrill when Sleepy Cat Farm is open. Sometimes he likes to fade into the background and act as docent or ticket taker, politely answering questions about the owner and garden maker. "It's wonderful to be able to bring joy to folks and to hear their comments. I've never seen people leave with a scowl on their faces." Fred Landman confirmed that the most important thing to him was that people tour his garden and leave happy. That's certainly how I felt on both my visits to Sleepy Cat Farm.

Ken Druse

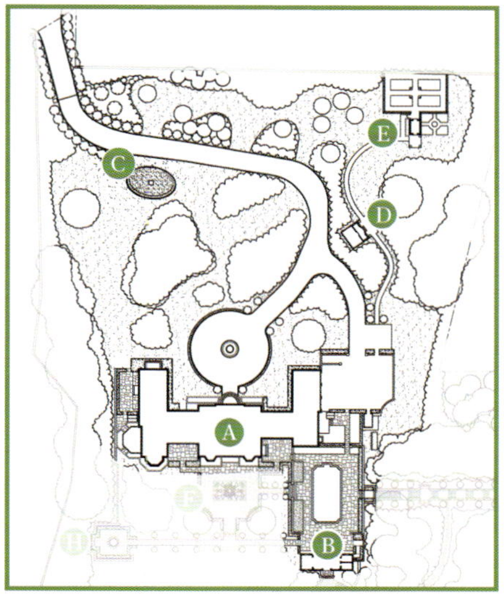

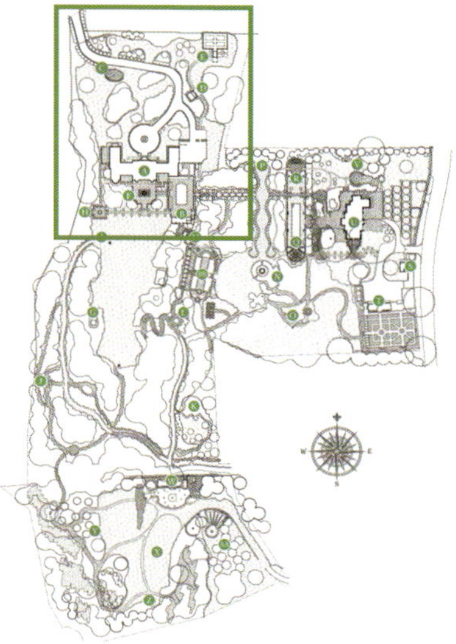

THE HOUSE

"I wanted to make sure I was creating something beautiful."

—FRED LANDMAN

When making a garden we must, as all the great garden designers tell us, start with the house. In a secluded residential area of Greenwich stands a neo-Georgian house completed in 1942. Made of red brick, with a pedimented projecting facade, columns on either side of the front door, a gable, dormers, high chimneys, and Palladian windows, it was purchased by Fred Landman in 1994.

Landman had traveled widely on business over the years, and he always admired the historic architecture of European cities, particularly the Georgian terraces of Dublin and London. With the rich stock of Georgian revival houses in Greenwich, he was particularly drawn to that style. "I loved the elegant details," he says. "I loved the brick facade."

It was a perfectly good house, but Fred Landman wanted more. He was on his own, his children were grown, and he was looking at the house as an important part of his new life. "I knew I wanted to update it," he says. "It was solidly built and had great bones, but there were problems. And as I was starting over, I wanted to make sure I was creating something beautiful."

Landman selected Charles Hilton (known as Chuck), a young, highly regarded classicist architect, to work with him to transform the house into something closer to his ideal. Hilton had a strong background in historic architecture, in particular the Georgian architecture Fred Landman so admired. "When my kids were small, we would drive to South Carolina to visit my in-laws in Hilton Head," Hilton explained. "I would always try to work in an overnight and visit to some plantation houses each spring."

Preceding pages: Aerial view of the house, entry court, and pool area.

Initially, Landman's ambitions were modest. He identified the main issues as arising from the interior of the house, and specifically from the kitchen, which he described as being dominated by "a pink Formica table and a green banquette." Hilton's first assignment was a complete kitchen renovation. But it was not long before the two men realized they had much more to talk about than countertops.

As Hilton began working on the house, later called Sleepy Cat Farm in honor of the many felines who moved in over the years, it was soon clear that it needed more than a kitchen upgrade. "The house had not received attention for many years," he commented. Gradually he transformed the other rooms, changing lines of axis, adding light, creating flow. Upstairs, he updated bedrooms and bathrooms. Interior designer Isabelle Vanneck of Davenport North, was brought in as part of the team to work on the decoration.

Meanwhile, Landman had been looking at the land in front of the house. Here he took his first gardening steps, planting a small vegetable garden and some flowers and building a neoclassical tool shed. Chuck Hilton was looking out from inside the house

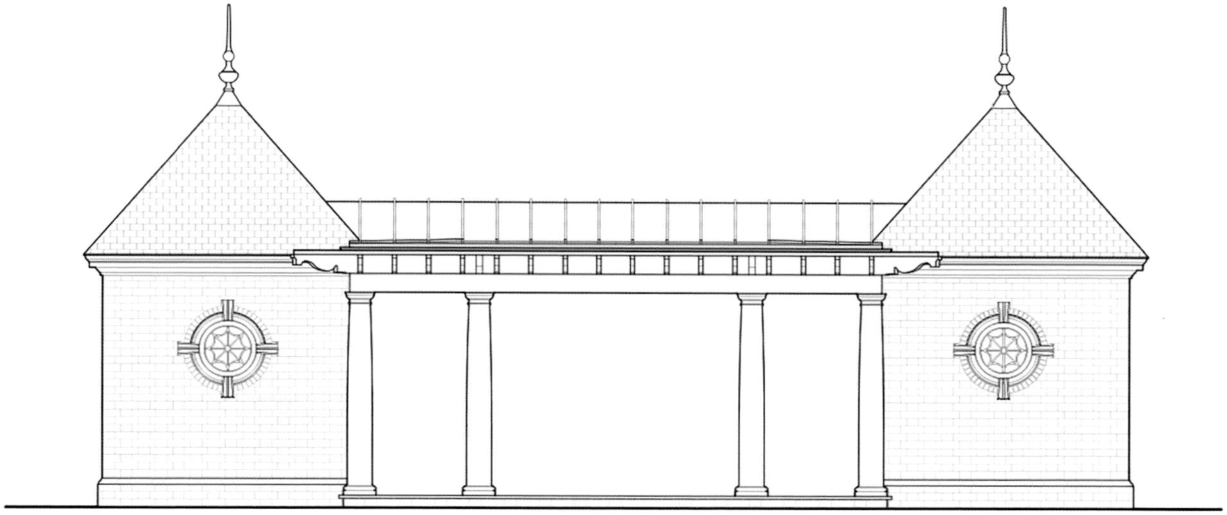

Above: North elevation of the pool pavilion.

to the exterior. As he continued his conversations with his client, more and more aspects of the house became subject to scrutiny. The first element of their critical review was the pool-house area. There was already a swimming pool, installed on the east side of the property. It was not a beautiful thing. Neither was the pavilion. "Very small and quirky," Hilton observed. Landman agreed. "I wanted a place to change, prepare food and drink and dine poolside," he said. The challenge was the limited amount of land available for reconfiguring the pool house.

As discussions continued, by a stroke of luck (one of many that have blessed him over the years), a neighbor offered to sell Landman a quarter acre of land at the east side of the pool area. With this extra space, Hilton was able to design and build a truly elegant pool house. It has as its core an open pergola dominated by four slender white columns and flanked by two brick pavilions with hipped roofs and oculus windows. Between the pavilions is a dining terrace; inside are changing rooms, a kitchen, and a bathroom. At the rear is a trellis that acts as a backdrop and also defines a fairly steep slope behind the building, where an outdoor cooking area was added, and steps were built leading down to the main part of the garden. The footprint of the pool was retained and enhanced by a wide stone surround for chairs, tables, and umbrellas, and, perhaps more important, a terrace.

Left: An oeil de boeuf window in the pool pavilion, surrounded by climbing *Clematis montana* 'Rubens'.

Right: Versailles boxes planted with Limelight *Hydrangea paniculata* 'Grandiflora' punctuate a yew hedge beside the pool.

A terrace requires landscaping, and for Fred Landman and Chuck Hilton, it was a turning-point. To pull off this important aspect of the reimagined pool area, Landman realized he needed a new expertise. He found it in Charles J. Stick, a landscape architect from Charlottesville, Virginia. Landman saw Stick's work in a neighboring Greenwich garden and admired his treatment of the pool terrace and its connections to the main house. Charles Stick came with fine credentials, including a degree in landscape architecture from the University of Virginia, where he found a lasting love of Thomas Jefferson and Palladian architecture, and later collecting a handful of ardent, loyal clients. He arrived at Landman's property in early 1997, and, like Hilton, he struck up a conversation with Landman that has turned into a more than twenty-year dialogue.

Left: The bed on the west side of the pool is planted with *Hydrangea paniculata* 'Tardiva'.

Right above: Versailles planters flank the steps from the pool terrace to the parterre.

Right: Climbing *Clematis* x. *jackmanii* softens the formality of a brick pier topped with a lead Adam-style urn.

Above: Entrance and garden facades of the house before the west wing addition.

The landscaping around the pool area was Stick's first contribution to Landman's garden, adding a bluestone path and staircase from the house to the pool as well. But this was just a preface. The project was an early success of what became an extensive program of renovation and building that continues to this day. Encouraged by the pool area expansion, Fred Landman was now moving into high gear, aiming the spotlight on every aspect of the house and outbuildings. Not that things happened randomly, or overnight. The developments were incremental as the team, now with Stick as a member of the trio, moved from one part of the property to another.

The most obvious place that required attention was the entrance to the house itself—the arrival point, the parking, the front door, what visitors would see first. This was the most significant issue that occupied Landman, Hilton, and Stick during these first years of exterior renovation. The main issue was the driveway, which was blocked by a large old oak tree and could not complete its circular course to the entrance. After some soul-searching, Landman had the tree removed, thus allowing the driveway to complete its journey.

While the driveway was being reconfigured, Chuck Hilton was already focusing on the "weaknesses" as he called them, of the portico and cupolas of the entrance facade. Working with Hilton, Charles Stick created a new planting area around the circular court defined with large boxwood and a low yew hedge. Soon the front entrance, driveway, and parking were making a much stronger first impression.

Above: The renovation included creating a welcoming entry court and adding a new wing to provide the characteristic symmetry of traditional Georgian houses.

But as Fred Landman stood outside the house and looked at it, he was troubled by a lack of symmetry, an essential element of Georgian architecture. The house "had an unbalanced appearance, with a large east wing jutting out from the main core of the facade." The solution: a west wing that would create the balance of the Georgian ideal.

With Landman's enthusiastic support, Hilton replaced a screened porch with a wing that would provide splendid new spaces that Landman realized he wanted—a great room, a garden room, and a two-bedroom guest suite. Now the house has a symmetry, a finality about it that was lacking. In the garden room, with its tall arched windows opening on to the back parterre and shimmering murals of flowers from the garden by Robert Kushner, an old friend of Landman's, there is a profound sense of connection between the indoors and the outdoors.

Please join us in our celebration of marriage!

Seen Lippert

&

Fred Landman

Saturday, June 21, 2003

Ceremony begins at five in the evening

Garden Reception & Amazing Feast to follow

146 Clapboard Ridge Road

Greenwich, Connecticut

The completion of the wing in 2003 might be said to mark the end of the first period of Fred Landman's architectural interventions. "The day the addition was completed, Seen Lippert and I were married at the house with 150 people in attendance," Landman recalled. This happy celebration served not only as a spectacular affirmation of years of work well done, but also as a window into the future. The renovated interior, the new pool house and landscaping, the path and staircase connecting the house to the pool area, the new west wing—all these improvements were already beginning to inspire Landman to look even further afield.

Above left: Eighteenth-century marble figures of Paris and Helen face each other across the driveway.

Above right: Invitation to the wedding celebration of Seen Lippert and Fred Landman at Sleepy Cat Farm.

Above left: *Apollino* by Thomas Crawford (1837) and *Glancing at Nymph After the Bath* by Carl Cauer (1870).

Above right: Umbrellas shade tree peonies to extend bloom time.

A classically inspired tool shed graces the upper garden. *Hydrangea macrophylla* 'Blue Wave' and Asiatic lily 'Alba' are among the border perennials.

Plantings in the upper garden include Azalea 'Northern Lights', *Hydrangea arborescens* 'Annabelle', *Viburnum plicatum* 'Mariesii', *Digitalis excelsior* 'Alba', and English bluebells.

The upper garden includes the original vegetable garden and a small greenhouse where vegetables are started in the spring and house plants are revived in the winter.

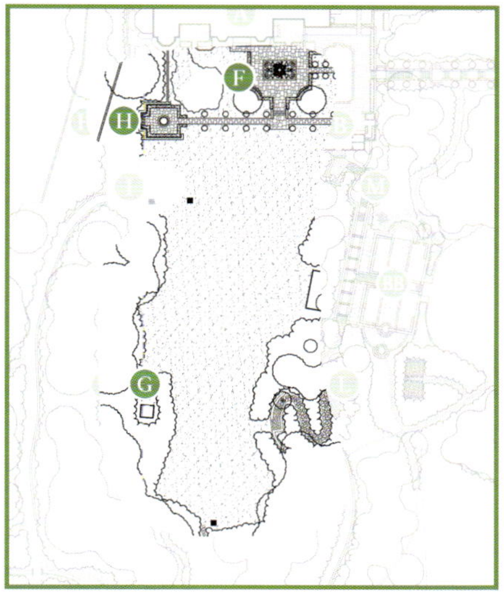

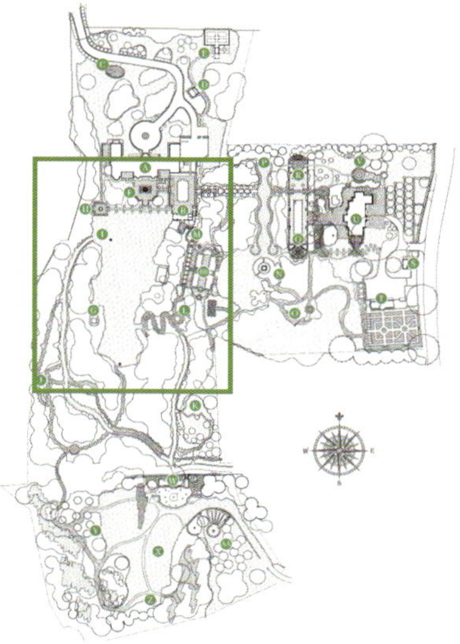

THE LAWN AND THE RIDGE

"Our mission was to create a garden of which the house could be proud."

—FRED LANDMAN

Fred Landman began to realize that with a re-imagined house, he should also have a garden that would reflect the house's elegant form. He understood that all the great estates he admired so much in Europe were inextricably connected in one specific way—the house with the garden. As the distinguished British garden designer Sir David Hicks observed, "Notice how old houses and gardens are always beautifully related. A good and successful garden, however new, will always have this sense of harmony."

By the late twentieth century, the leading opinions about garden design had progressed into two main camps—the formal and the formless. Put another way, the theory behind garden design leaned either towards the man-made or the naturalistic. Lancelot "Capability" Brown, the eighteenth-century landscape gardener, said, a propos of an English garden that he was pleased with, that it contained "facile compositions of grass, tree clumps, and rather shapeless ponds." Brown and his followers liked the natural look of a garden, emerging as though by chance and nature's whims. Two centuries later, the gardener Russell Page concluded that "the formlessness of so many modern gardens stems from this earlier decadence." For Page, the designer imposed his will on the landscape to ensure its harmony of line and shape. The differing styles expressed by these two geniuses were to dominate much of Western garden design in the twentieth century.

For Charles Stick, the design of a garden should first and foremost be on a human scale, and it should combine the useful and the enjoyable. As he says, "Order in itself provides a fair amount of pleasure to the spirit." Thus does Stick's work express both the formal and the naturalistic in a happy truce.

From his point of view, the first issue for Landman's new garden was the outstanding problem at the back of the house. With the pool landscaping resolved, he had to address the fatal flaw as he saw it—there was no transition between the house and the lawn. "As frequently happens," Stick explains, "so much money and effort are spent on the house that the land immediately surrounding it is ignored. We needed some resolution to make Fred's house be received by the land."

Stick's solution was to build a parterre, a French term originally used to describe a symmetrically planned herb garden connected by paths and later developed as a series of beds planted in elaborate "embroideries," usually of boxwood, and placed close to a palace or mansion so the effect could be viewed to its advantage from the formal rooms of the residence. Stick's parterre for Landman followed the French pattern. It was surrounded by a low boxwood hedge that bordered the central enclosure containing an elaborate pattern, like a carpet, ideal for viewing from above, in a layout that evokes the formal gardens of England, Italy, and France. This decorative space unfolded immediately outside the back door of the house, creating a pleasing physical and visual relationship between house and garden, as David Hicks recommended.

Preceding pages: The view from the principal suite extends over the boxwood parterre and the lawn to the Celestial Pavilion. *Clematis montana* 'Rubens' climbs up an iron trellis to the left, adjacent to one of two small leaf linden trees (*Tilia cordata* 'Green Spire').

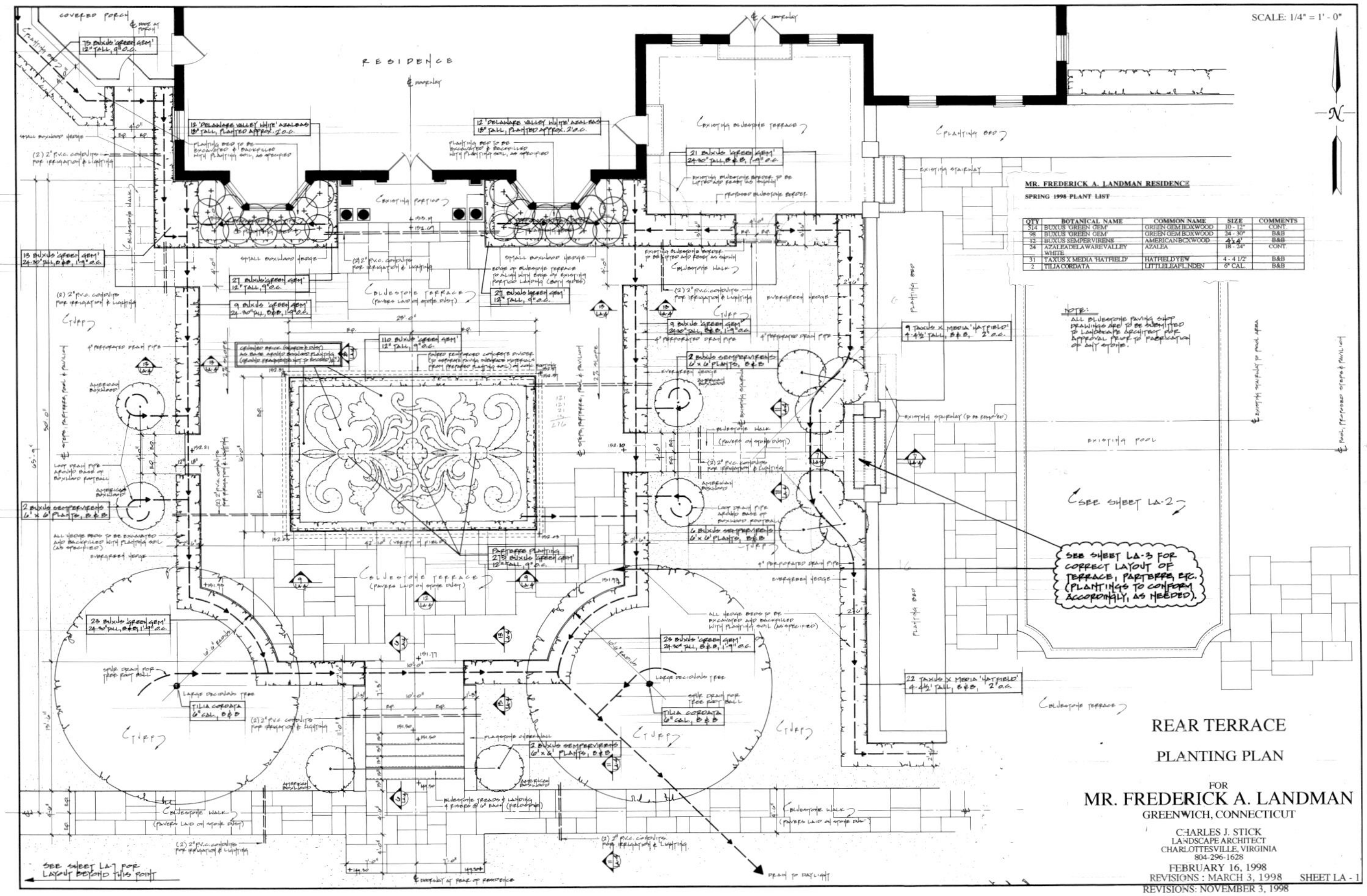

Above: Planting plan for the south terrace, February 1998.

Stick's original idea for this plot also included a rose garden, which he situated on the west side, adjacent to the path to the pool area, creating two cross-axes. But at that time Fred Landman did not find roses particularly appealing, so Stick placed a fountain there instead. "Those were the initial strokes," he said. "All architectural additions close to the house."

But something happened during this transformation of the back lawn. Fred Landman saw that the parterre in his garden in Greenwich bore resemblances to the formal designs of those in England and France, and he loved it. He thought about it, and he began to sense that the rest of his undeveloped garden had

Above: The landscape behind the house was a forlorn piece of grassland and rocky outcroppings, with a pavilion constructed in plastic, before the installation of the parterre and lawn. Daffodils now soften the rocks in the spring.

significant potential. "I wanted to make the outside as inviting as the inside," he said. And his question to Charles Stick was "What shall we do next?"

The two men spent days, weeks, months, contemplating the long sweep of lawn and rocks, surrounded by woodland, that are the constituents of the Landman landscape. They gazed at the ridge, the large outcropping of ancient granite and gneiss that runs from north to south through the middle and west side of the property, like the spine of a sleeping dinosaur.

Charles Stick studied the lawn, the rock ledges, and the surrounding forests. To him the scene had the qualities of a park—a mixture of light and shade, trees and grass. Formless, yes. "It had no openings at all," he recalled. "It was all overgrown, seriously neglected. There was no sign of form."

Stick knew about form. He had seen formal gardens throughout Italy, France, and England. He had designed formal gardens at Waverley Farm, his home in Charlottesville. He had encouraged Fred Landman to see Vaux-le-Vicomte, the most extreme expression of French formal design garden in the world. His first move in Greenwich was to give a focal point to the lawn sweeping southwards with a newly

Right and below: The Celestial Pavilion, designed by Charles Hilton, replaced the plastic structure. A view of the constellations, inspired by a sixteenth-century astrological woodcut, is incised into the black marble floor. The elegant brick structure received a design award from the Connecticut chapter of the American Institute of Architects in 2008.

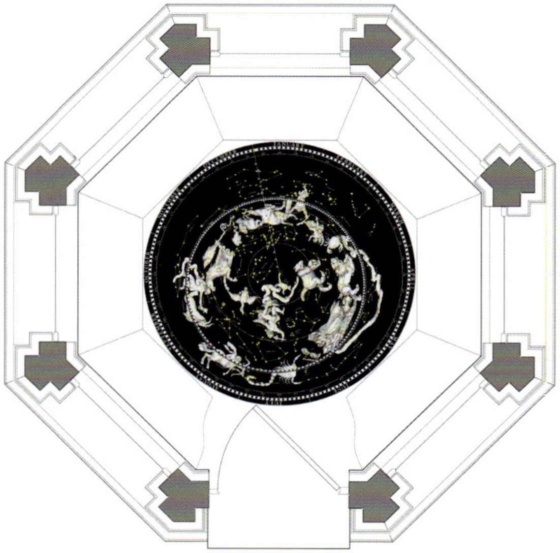

Left: Two iron trellises entwined with *Clematis montana* 'Rubens' give a vertical lift to the boxwood shapes on the parterre.

Right: A mid-nineteenth-century sundial has four faces that turn toward the sun depending on the season.

Above: A copper roof protects the hexagonal purple martin house.

Right: Aristide Maillol's *L'Été sans bras* stands out against the rich fall color of the woodland.

Left: As the woodland is cleared of invasive plants and the understory is ready to be planted, the Golden Path takes shape.

Opposite: A secondary rustic stone path lined with spring bulbs connects the Golden Path to the rear lawn.

designed pavilion on a little rise. It came to be called the Celestial Pavilion, and it was designed by Chuck Hilton, who was full of ideas for further enhancement of the property based on American architecture from Colonial Williamsburg to the James River plantation houses.

Charles Stick became absorbed by another, more challenging preoccupation—the ridge itself. Many a house-owner looking out at such a ridge would immediately order it buried, removed, or somehow concealed as an eyesore. But Fred Landman and Charles Stick had something else in mind. For them, the ridge was a gift—a work of nature that they could turn into an adventure.

Exploring the wooded areas of the ridge, they came upon a series of slabs, mounds, boulders, and glacial rock formations ("elephants' backs," as someone called them)

that made strolling through the landscape a journey of constantly shifting scenery. Far from getting rid of these ancient geological strata, they wanted to embrace them. "There were foxes, deer, wild turkeys in the woodlands," Fred Landman remembers. "We wanted to preserve that feeling of wildness." Charles Stick agreed. "The topography tells you what to do," he says. "We just had to follow where it took us."

With these simple guidelines, Landman's garden evolved in dramatic fashion. "In designing the layout, always approach it in a spirit of adventure, enjoyment, and also patience," advised Sir Roy Strong, former director of the Victoria & Albert Museum and himself the creator of a delightfully personal garden in England. "Be bold and never be put off by gloom-mongering friends who tell you that you will never live to see the fruits of your creation. Even the making of the garden will give untold pleasure."

Stick's earliest drawing of the garden is like an unfinished manuscript. Dated October 8, 1997 (with additions a year later), it shows the garden in its infancy. The Celestial Pavilion and other architectural touches are identified, along with the parterre and pool area. The other notations he makes include the cooking area at the back of the pool house and, a little further south, an azalea and hydrangea garden. The overall plan is dominated by the carefully drawn wooded areas surrounding the perimeter of the landscape. "Existing woodland," Stick writes to describe the shapes and circles crowding together around the boundaries of the garden.

Above: Fred Landman places plants along the Golden Path.

Opposite: Original 1997 master plan by Charles Stick for the rear terrace and pool, with suggestions of things to come.

Later plans show many more detailed architectural elements that were introduced as gradually the more formal areas to the east were developed. But for these first years, it was all about the ridge, uncompromisingly formless. "You're not getting a big flat croquet lawn here," Fred Landman asserted. "The woodland, the trees, and the rocks created an opportunity to design varied garden experiences." The ridge allowed Stick and Landman to solve the issue of formality versus informality that has continued to divide garden designers. Today, Fred Landman's perennial border, Chinese Pavilion, reflecting pool, steep stairways, undulating hedges, boxwood balls, and many other features from the landscape architect's toolbox offer the visitor formal gardens straight out of the European tradition. But those who walk across the lawn enter another world entirely—jagged rock formations, uneven stone paths, swathes of colorful ground cover, hillsides dotted with mysterious statuary, every vista shaded by dark canopies of trees.

From nature's woodland to man-made beds and borders, from the wild to the tamed—making these connections is the basic story of the gardens at Sleepy Cat Farm And the ridge on Landman's property, that sleeping dinosaur, proved to be the catalyst that turned the landscape into a paradise of contrasting landscape styles.

LANDSCAPE MASTER PLAN
FOR
MR. FREDERICK A. LANDMAN
GREENWICH, CONNECTICUT

CHARLES J. STICK
LANDSCAPE ARCHITECT
CHARLOTTESVILLE, VIRGINIA
804-296-1628

OCTOBER 8, 1997

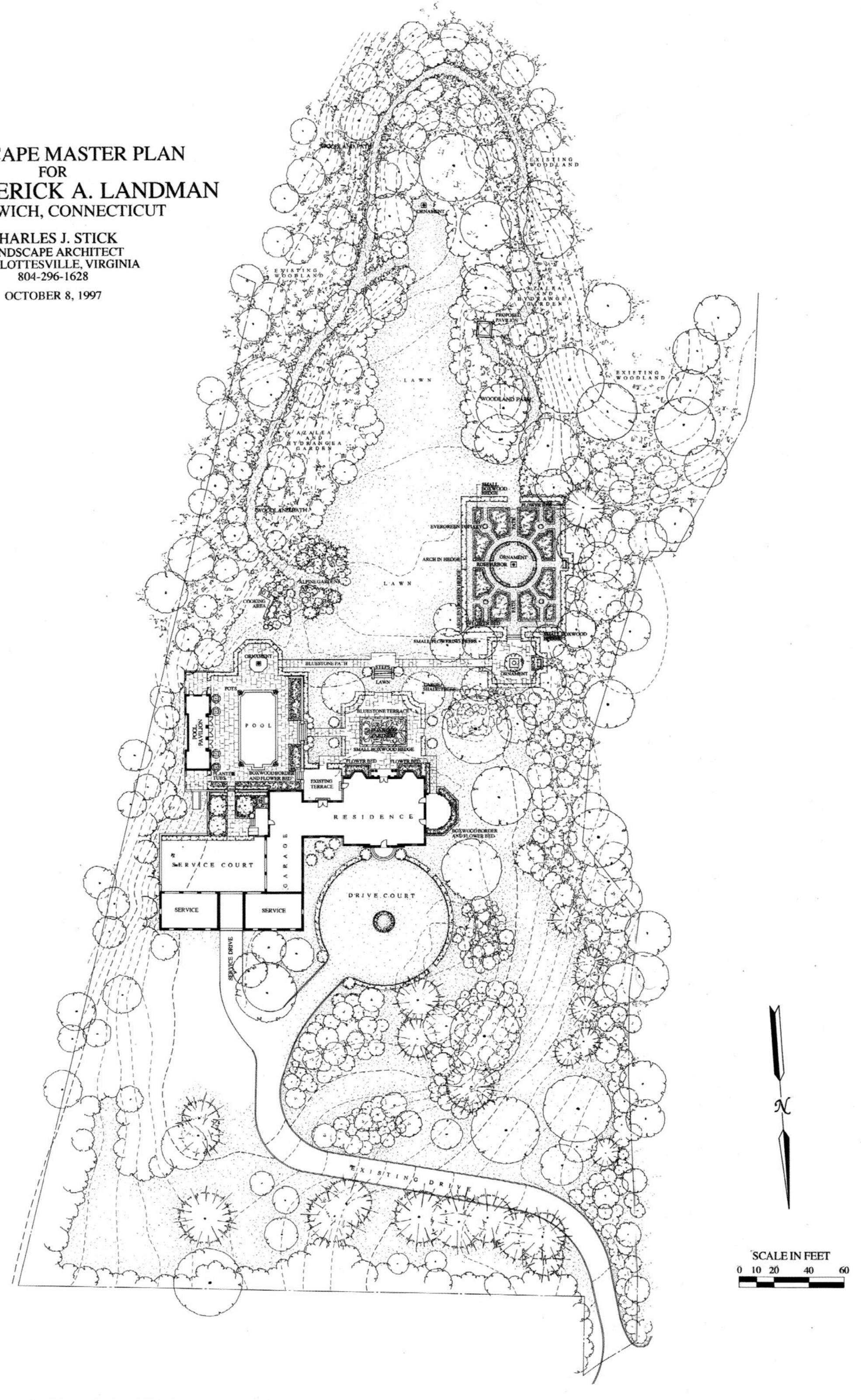

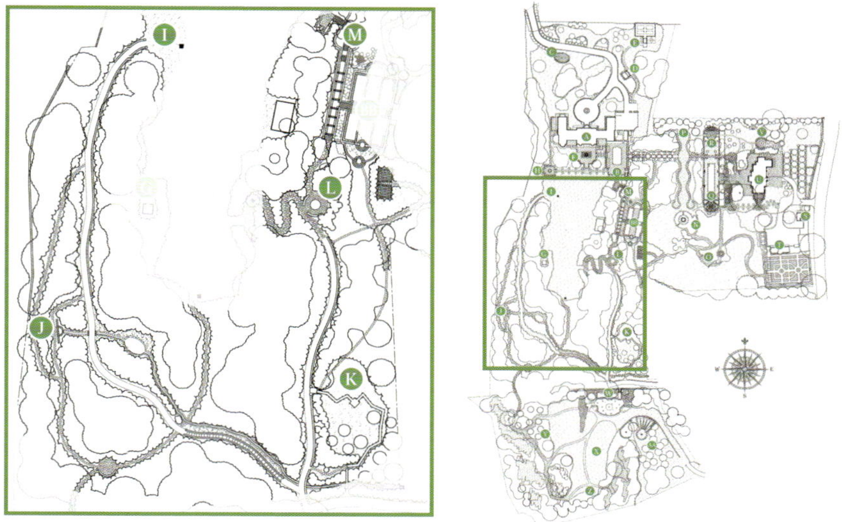

THE GOLDEN PATH

"You could be dropped anywhere in the garden and always feel the thread of the narrative."

—CHARLES STICK

From the start, Fred Landman and Charles Stick had a similar vision, one that Stick articulated in talking about his own garden in Virginia. "The idea was to give someone moving through the landscape a sense of discovery," he said, "to create layers of unfolding experience."

Charles Stick's idea originated in the work of Russell Page, who designed gardens all over the world, both public and private. His last project, in 1980, was the world headquarters of Pepsico, Inc., a one-hundred-acre site in Westchester County, New York. His challenge was to harmonize the company's twentieth-century sculpture collection with the extensive parkland in which the large-scale works were installed.

Page's solution was to design a path made of amber-colored gravel that curled round the property, linking the offices, sculptures, and garden areas together in a vast embrace. In notes he wrote at the time, he mentions the problem of "fluidity" in the original composition of the landscape, how to enhance the experience of the visitor walking through the grounds, just as Charles Stick was to grapple with at Sleepy Cat a decade later. Page's winding gravel path was almost like a river flowing through the landscape, unifying the separate elements. This feature became known as The Golden Path.

Charles Stick's 1997 plan includes a design feature that at that time was simply identified as a "woodland path." It runs from the west side down through the wooded areas to the end of the property, then curls round and up the other side toward the pool. Today it is clear what that modest drawing was all about.

The challenge was not unfamiliar to Stick. After Russell Page's death, Stick had worked for six-and-one-half years at Pepsico, implementing Page's plan to unify the disparate elements of the Pepsico sculpture gardens. With Fred Landman, it was up to Stick to find the secret that would unlock the much smaller but equally diverse landscape spread out before them. It wasn't long before they found the key—an animal trail that led off to the right of the property. "If the deer had found a good way to move through the woods, didn't it make sense for us to follow their lead?" Fred Landman asked. Answering this question was the beginning of Sleepy Cat Farm's Golden Path.

Landman and Stick had only six-and-one-half acres to work with, rather than Russell Page's more than one hundred. And while Page had to incorporate twenty-eight sculptures, Landman and Stick had no sculpture at all at the start of their project. The first one, an Aristide Maillol entitled *L'Été sans bras (Summer without Arms)*, was bought by Landman when the Golden Path was completed. Landman's landscape turned out to contain a treasure trove of inspiration, tucked away in the

Preceding pages: The entrance to the Golden Path is marked by a pair of cast-iron urns filled with Boston ferns.

Above: In spring, white azaleas animate the Golden Path; a stone path leads through a display of daffodils to a welcoming bench at the crest of the ridge.

jungle of trees, rocks, shrubs, plants, and weeds, treasures that became the essential elements of their Golden Path. "The ridge created its own series of rooms," Stick explains. "It was a matter of tying everything together."

The path offers itself without much fanfare. On the visitor's map it is simply a small opening at the west side of the property, hardly noticeable when one is absorbing the formal parterre, boxwood and island-studded lawn stretching down from the back of the house towards the Celestial Pavilion. But stepping through the "gateway" marked by two urns planted with ferns and mounted on pedestals, visitors are immediately transported into very different terrain. The land on each side becomes uneven, shaded, marked with rocks, trees, wildflowers, and shrubs. The simple decomposed-granite trail, defined in part by low stone walls, begins to wind through surprising rock formations and dense woodland. It is clear that this is going to be more than an ordinary stroll—it will be an adventure.

Left: The Grotto basin, another artful experience of the Golden Path, with sweeps of hosta and ferns.

Above: *Rhododenron viscosum* and bleeding hearts in dappled light near the Grotto.

From the beginning, Charles Stick had a strong sense of what the Golden Path could do. "As we began to lay out the path, we wanted to tap into the constant presence of the ridge," he explains. "The ridge was our lodestar—it was the genius of the garden." Indeed, the ridge is never far from the Golden Path. It looms, hides, rises and falls, takes shape in a constantly unfolding panorama. The shade is at times intense. A vast canopy of trees makes dark shadows fall across the rocks and undergrowth. Some of the ancient tree trunks seem contorted and gnarled like gnomes from *The Lord of the Rings*, or perhaps the evocative illustrations of the Victorian fairy tale illustrator Arthur Rackham.

Above: The entrance to Valsanzibio, a garden near Venice, the source of many ideas for Sleepy Cat Farm.

In one of their early conversations, Stick had suggested that Landman visit Valsanzibio, a glorious seventeenth-century garden near Venice. That garden boasts three grottoes, and Landman loved all of them. When he got home, as he walked through the looming rocks and shadowy trees, he asked Stick if they too could make a grotto.

The Sleepy Cat Grotto is the first man-made structure along the Golden Path and it comes into view with elegant stealth. Grottoes are natural or artificial caves, often with water cascading out of their openings, nestled in rocks or hillsides. A popular feature of many landscapes dating from Greek and Roman times, they represent an aspect of garden design sometimes underestimated by today's garden designers—the notion of playfulness. They are appropriately called "follies."

Landman and Stick's Grotto is strategically placed at the foot of a powerful stone mass, rising up steeply from the path like a small mountain, an example of Stick and Landman's respect for the natural rock formations that inspire the architecture along the Golden Path. The structure is limestone, made to Stick's specifications, with an arched opening flanked by two columns with lead urns (their finials worn down by chipmunks sharpening their teeth, according to Landman), and stepped walls on each side merging into the hillside. The rushing sound is that of a waterfall, emerging from the mouth of the cave, bubbling down a narrow channel

Right: The construction of the Grotto, showing the large stone outcropping that became the perfect backdrop for the feature and stakes laid out by Charles Stick to mark the course of the stream leading to the Grotto basin.

decorated with stones and moss. This waterway runs under the path and trickles gently into a reflecting pool, where a statue of a crouching Venus admires herself in the water.

The steep slope above the Grotto climbs to a plateau, where a welcome bench is waiting. "When people are hiking uphill, they should always have a rewarding experience when they get there," Fred Landman observes. From this spectacular vantage point, you can look down over the Grotto, past the pond to Venus and the surrounding trees, flowers, and woodland that are so much a trademark of the

Opposite: The Grotto and waterfall, planted with ostrich ferns, flame azaleas, and *Helleborus orientalis*.

Above: Moisture at the Grotto spillway creates a bed of moss for frogs to play in.

Above right: A twentieth-century limestone *Crouching Venus* at the edge of the Grotto basin.

Sleepy Cat garden experience. In a typical Landman touch, from time to time puffs of mist emerge from the Grotto, further entertaining the visitors admiring the scene.

The Sleepy Cat Grotto looks more English than Italian. In fact, for those familiar with English gardens, it looks remarkably like the grotto at Rousham House in Oxfordshire, where Charles Stick first saw and was smitten by it. The grotto at Rousham, which has less ornament and no canal, and emerges from a flat rather than a rocky backdrop, was designed by the eighteenth-century architect William Kent. His work at Rousham has always been "a great inspiration to me throughout my years as a landscape architect," in the words of Charles Stick.

It wasn't until many years later that Fred Landman saw for himself how closely his grotto was related to the one in England. He traveled to Rousham and stood before Kent's grotto and saluted the work that had provided such inspiration for a garden in the New World. (Incidentally, neither Landman nor Stick makes any apologies for this adaptation. Stick calls it "precedent.") This collaborative way of doing things would form a template for almost all the garden design features of Sleepy Cat Farm to come.

Moving on, the Golden Path winds its way through swathes of understory planting (in the spring, countless bulbs), woodland flowers, and thinned-out stands of beech, oak, and linden trees that produce a constantly changing play of light and shade. Along the way, a stepping-stone pathway covered in moss appears on the left, leading up a rocky slope away from the garden path. Where is it going? Where has it come from? The steps in fact can be traced back up to the rear lawn of the house, offering

Above left: William Kent's Vale of Venus at Rousham in England, the inspiration for the Grotto at Sleepy Cat Farm.

Above right: Rendering of the Grotto.

Opposite: Every spring an explosion of vivid blue camassia lines a seasonal stream.

secondary access to and from the Golden Path. There are several auxiliary paths like this, some still being developed, offering up unexpected vistas and openings for the visitor to explore. In and out of shadows, these stone stairs at times become scary as they wind up the leafy hillsides, past fierce rock formations and ancient twisted trees, beckoning into the unknown. "Everywhere I look, I want to see something that has me saying Wow!" Landman says.

The Golden Path then departs from this dark and tangled landscape and heads inland, so to speak, turning east and on up northward toward a whole new flow of experiences. Originally, the path turned east here because the property ended roughly at this point. The rest of the garden stretched back toward the house on the other side of the lawn from where the path originated. Over the years, Landman was able to acquire three-and-one-half more acres to the south, now reached by another series of trails. The Golden Path, meanwhile, retains its role as the main unifier of the property, though now offering very different views of that precinct, which by this time includes the Limonaia and its sunfilled terrace.

From this point, the path winds back toward the house, away from the shade and into the light, with invitations to linger on both the left and the right. On the right is a raised walkway made of Spanish cedar, known as the Spirit Walk. It zigzags over a wetland plain, planted, after some experimentation, with a mass of Japanese irises. The sharp angles of the walkway are believed by the Chinese to discourage evil spirits. At the far end, a stone head of the Buddha reinforces the idea that this is one of several areas designed to reflect the spiritual subtext of the garden.

Right: The Spirit Walk, made of Spanish cedar, is enveloped in *Iris ensata* 'Lion King'.

Opposite above: A Buddha head from India is mounted on a rustic stone column beside the Bamboo Walk.

Opposite below: The spring bulbs of *Anemone ranunculoides* make a carpet of yellow.

The path of the Spirit Walk changes direction as it traverses the Iris Garden, reflecting the Chinese belief that sharp angles discourage evil spirits.

Above left: The dark gray urn and pedestal, c. 1840, stand out against vivid azalea blooms.

Above right: The Golden Path leads to a circular pool where a figure of Narcissus is installed at the edge.

On the left side of the path, a spiral column made of warm Bath stone, topped by an urn, rises up from a steep rocky hillside. This unexpectedly wild scene makes a dramatic contrast to the serene, lake-like wetland on the opposite side. Moving up towards the main house, the path passes stepping stones that lead to the Bamboo Walk. Here the temptation to venture even further east, both literally and metaphorically, is provoked by the sight of three sets of Indian prayer wheels.

But the Golden Path turns left, beckoning the walker northwards, past a basin with a single jet and a statue of Narcissus. It continues through the forest of hydrangeas as marked on Stick's 1997 plan. To the right are tantalizing glimpses of the colorful perennial border. The path finally comes to a steep flight of rustic steps to Mercury's Arbor. Here a statue of Mercury stands poised in welcome as the Golden Path winds to a close, tying together the various garden "events," as Charles Stick calls them, that have been placed along the way. "We wanted it to have a sense of narrative," he says.

That narrative is not yet finished. Today, the garden has doubled its size since the Golden Path was first conceived, and the collection of sculptures has risen to more than twenty and is still growing. The Golden Path remains a haunting passage through the old and the new, the driving force binding together its ancient and modern history.

Russell Page wrote about his Golden Path that "the nature of the sculptures and their materials demands contrast in plant forms and textures and a careful study of light and shadow." These could be the words of Landman and Stick describing their own personal version. "I walk around it at least once a day," Fred Landman says. "Every time something changes, something looks different. Every time, I ask, 'How do we make this more interesting?'"

Above: Rendering of Mercury with stairs to the pool pavilion.

Left: Original wisteria arbor above the blooming perennial border.

Right: An Italian bronze figure of Mercury after Giambologna, mid-nineteenth-century, at the end of the Golden Path.

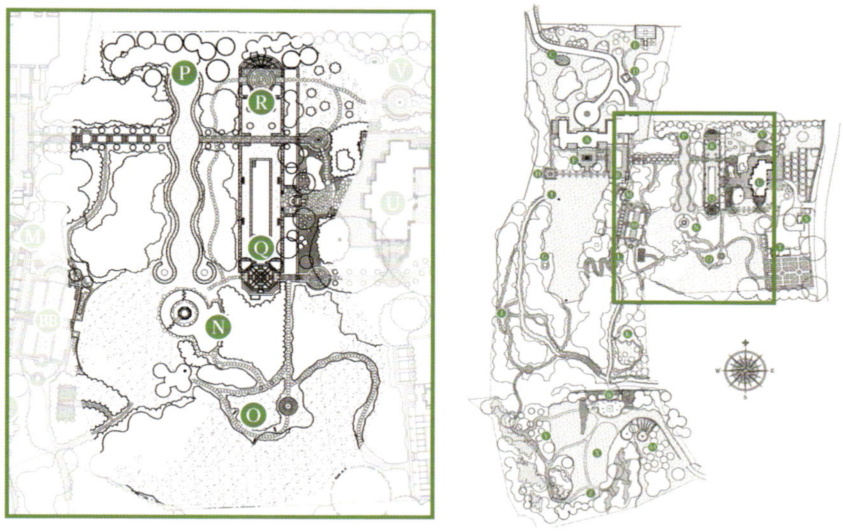

EXPANDING HORIZONS

"A pebble gets thrown in the pond, and everything radiates from that."

—CHARLES STICK

By the time the Golden Path was finished—or at least its final shape was established, Fred Landman himself had traveled a long way. He had visited gardens in Europe and Asia; he had learned how exciting and challenging garden design could be; and he knew he wanted to create something equally inspiring back home. He had seen the way Charles Stick took the Sleepy Cat landscape and unpacked it, testing its strength, learning its secrets. "I wanted to create a narrative," Stick once told him. "It's a matter of putting the different pieces together."

Fred Landman was now practiced in the intriguing language of garden architecture. He looked at the property he had already transformed and began to ask for more. He became "invested," as Stick put it. What was the next step? For most garden-makers, there is a house with a specific piece of land attached to it, and that is what they work with—that is the canvas. But Fred Landman had ambitions that went beyond his original six-and-one-half acres of land. He had already been lucky in being able to purchase more land near the pool. In another stroke of good fortune, in 2005 two more acres became available on the east side of the garden, where the Golden Path moved north through the hydrangea beds and up to the statue of Mercury.

The challenge was apparent immediately. The dominant feature of the landscape was a kidney-shaped swimming pool. "There it was, notched into the hillside," Charles Stick recalls, with little enthusiasm. Stick was also aware of the innovative work of Thomas Church, the celebrated landscape architect who designed some of the most elegant kidney-shaped pools in California in the 1940s and 1950s. "We looked at it," Stick said, "and determined how we might make it more regular. Maybe make it round."

The "round" idea summoned up a place known to both Landman and Stick: the famous "Rabbit Island" at Valsanzibio in Italy, a whimsical moated island where rabbits graze around a pavilion. "A beautiful piece of garden architecture," Stick said to his client. "Maybe we could do a rabbit island?" Landman, of course, was delighted with the idea, but after much discussion he realized the rabbit idea was a little too ambitious, so he reoriented the project to building a Ting, a Chinese pavilion in which to sit, reflect, and take in the views of the landscape.

Charles Stick liked the Chinese theme. "I'd been traveling around in England," he recalls, "and had stayed at Cliveden, not far from London. Cliveden has a beautiful garden with a Chinese pavilion surrounded by water." He also thought fondly of the Chinese garden at Biddulph Grange in Staffordshire.

Above: Rabbit Island at Valsanzibio and the Chinese Pavilion at Cliveden were inspirations for the Chinese Pavilion at Sleepy Cat Farm.

Opposite: Rendering for the Chinese Pavilion.

Preceding pages: A Chinese carved-stone hitching post, Qing period, stands at the northern edge of the eastern garden.

And so it came to pass, and the move east became literal as well as geographical. For Charles Stick, it made sense in a more profound way than just adding a new garden style to the Sleepy Cat mix. In the eighteenth century, an obsession with chinoiserie gripped artists and designers in England and Europe, and Stick realized that the Chinese theme meshed beautifully with the period style of Landman's house. "I had these English garden images in my mind," he says. "This little bit of history helped justify our Chinese theme. There's a thread there."

The Chinese garden rapidly took basic shape: a pavilion, surrounded by water. Stick, knowing his client's fondness for water, decided to animate it with a koi pond. A cascade added the mesmerizing sound of falling water to the scene, which is reached by a small stone-stepped trail leading from the Golden Path down an incline towards the woods. The trail, with bamboo railings on either side, became the Bamboo Walk.

Above Left: Grading the path between *Atlas* and the Chinese Pavilion.

Center: The pool for the Chinese Pavilion has an island in the center and large stone coping blocks around the perimeter.

Right: Charles Stick and Fred Landman confer on the direction of the pathway to the perennial border.

Opposite: The Chinese Pavilion, completed in 2001.

Left: Within the Chinese Pavilion are limestone melon-shaped garden stools, mid-Qing Period, and an inverted lotus-form marble relief carving, Ming Period, set as a table.

Right above: Umbrellas protect the tree peonies 'High Noon' at the koi pond cascade.

Right: A lotus mosaic by Robert Kushner is set in coping stone around the koi pond.

Left above: Japanese bronze cranes, Taiho Period, at the edge of the pond.

Left: One of eight koi that were introduced into the pond in 2001.

Right: The koi pond features both native and tropical waterlilies, along with lotus and flag iris.

Opposite: Magnolias and weeping cherry trees add spring color to the koi pond.

Above left: Early morning light illuminates the silhouette of the Chinese Pavilion.

Above right: A Japanese maple is perfectly framed by chinoiserie fretwork.

Following one of Fred Landman's visits to India, nine prayer wheels were added to the walkway, setting the mood for the spiritual elements to come. Landman had admired prayer wheels that dotted the streets in Leh, a town in the foothills of the Himalayas of northern India. With characteristic energy, he ordered three sets from a workshop nearby with no idea how he would get them home. They are now mounted on stands designed by Charles Stick and made in Charlottesville, Virginia.

Above left: A plywood mock-up of the prayer wheel stand was used to determine the final installation.

Above and opposite: The prayer wheels are nestled at the edge of a dense woodland, but the stand is placed on the pathway so visitors can spin the wheels as they walk by. Hydrangeas line the opposite side of the path.

Above: Past the prayer wheels, a red arch signals the entrance to a different garden theme. Spring bulbs 'Glory of the Snow' and Himalayan maidenhair ferns are planted along the path.

Above: A Han dynasty brick tablet, mounted on a stand for easy spinning, rises above the flowers of the woodland floor.

Above: A stone guardian crouches on top of an eighteenth-century hitching post from Northern China.

Right: A carved granite Kasuge lantern, Meiji Period, stands near the entrance to the Chinese Pavilion. Its hexagonal form and animated roofline echo that of the pavilion.

Opposite: Steps to the Japanese maple garden are lined with striking Japanese *niwaki*.

Above left: Chinese Foo Lion, nineteenth century, Fujian Province.

Above right: Guanyin, Hindu goddess of mercy.

The high point of this new extension was the Chinese Pavilion, or Ting. It has typical Chinese architectural features, curved roofs, a gold-leaf ceiling, and red-lacquer columns and fretwork. But the gold pineapple on top suggests something more Western. It seems to call to the finial on top of the Celestial Pavilion, which can be seen in the distance through the trees—folly to folly. Rising serenely from its watery foundation, the Chinese Pavilion has become one of the most popular destinations at Sleepy Cat Farm.

Left: A staircase incorporating a collection of French, English, American, and Chinese lion pairs, leads the steep slope to an obelisk.

Above and below: Laying out the foundation of the Lion Staircase and its sixty-five steps was a complex procedure.

Above: An aerial view of the obelisk shows its position at the terminus of the staircase cross axis.

The establishment of the Chinese Pavilion and koi pond set up a familiar problem for Charles Stick, one that he has confronted throughout his years working at Sleepy Cat Farm: how to connect one area to another. In this case, the Chinese garden was separated from the main house by a very steep slope. "How do we get people to the Ting?" he asked himself. The answer was a narrow stone staircase that would link the two areas. This became the Lion Staircase, lined with cats. At the foot of the staircase, Stick created a passage, or "hallway," as he called it, that would tempt the visitor toward the Ting.

The result was a grass-lined allée defined by two rows of undulating hornbeam hedges that terminate with a statue of Atlas at one end and the Ting at the other. "I remembered the serpentine hedge made by Joseph Paxton at Chatsworth," Stick explains. "It's a long, winding hedge, framing a green carpet of grass, casting shadows as it passes by a file of gardens on each side. We needed something like that to create a connection between the house and the Ting."

Above: The undulating hornbeam hedge at Chatsworth in Devonshire, England.

The conjunction of East and West suggested by the link between the Chinese garden and Atlas is not coincidental. Like the Ting and its pineapple, the East-West symbolism is a carefully thought-out design feature that provides the kind of literary flourish found throughout the garden that both Landman and Stick enjoy. "It's more than making something pretty," Stick says. "It's making something that has meaning, a context." Atlas became the metaphor for the West, holding the world on his shoulders.

The installation of the Atlas statue presented its own challenge. Sculpted by the Albanian-born, British-educated artist Andrian Melka, who works directly in stone, it is an enormous and extremely heavy work that had to be lowered by fork-lift on to a large plinth. Landman and his team were worried that the base of the statue would get scratched during this delicate operation so the house mason, Mauro Fidaleo, suggested using an old-world method of placing large blocks of ice on each corner of the plinth, and lowering the statue on to the ice. As the ice gradually melted, the statue slowly sank into place without damage.

When the Atlas garden was completed, it was clear that there was more land that could be developed to accommodate Fred Landman's desire to expand his garden. It presented another rectangular space parallel to the Atlas garden, and, again, Landman's love of water drove the plot. One day, while he and Stick were contemplating this new space, Stick said, "Why not create a water channel?" He was thinking in terms of a rill or canal with water jets like those found in the Moorish gardens in Spain and India.

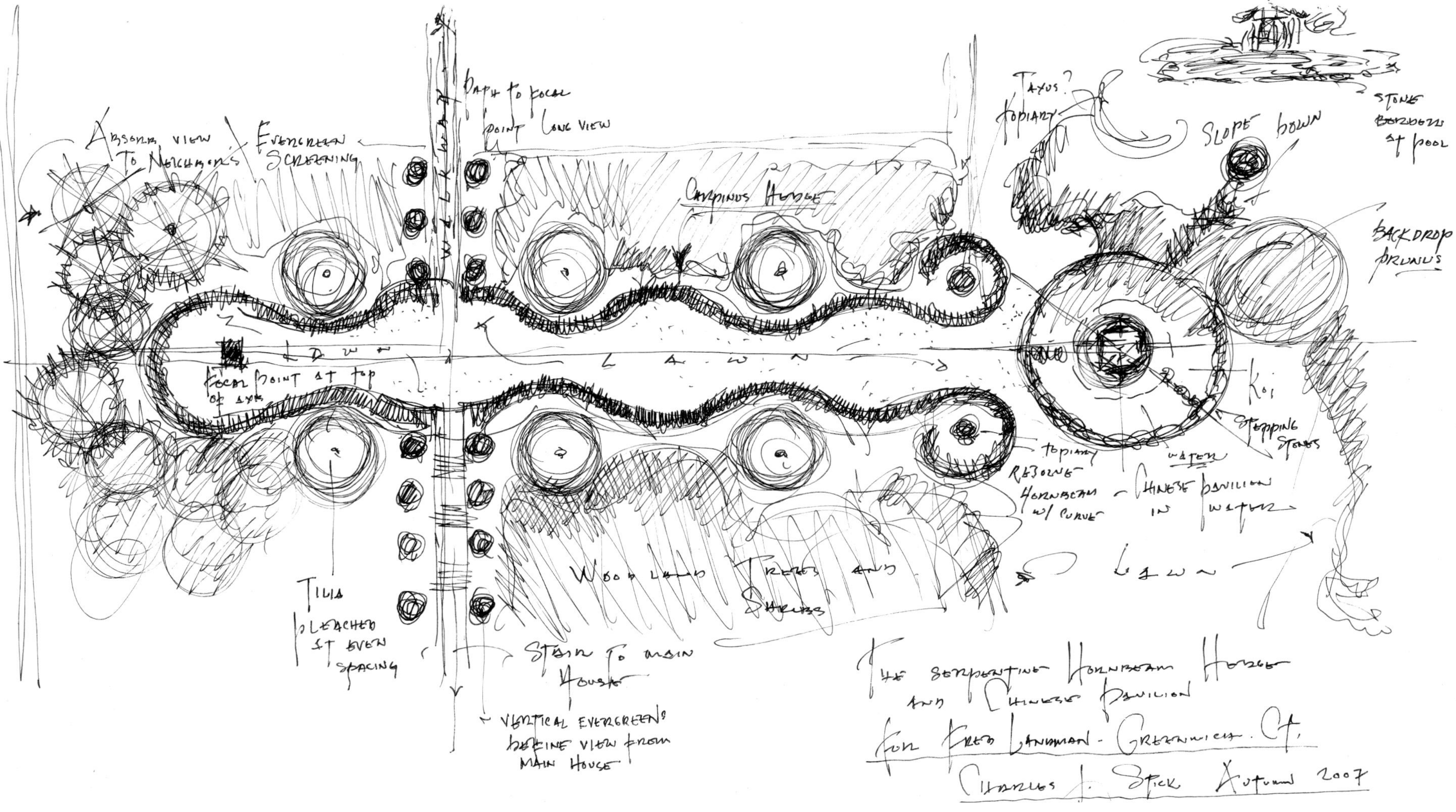

Above: Sketch by Charles Stick for the path between *Atlas* and the Chinese Pavilion bordered by a double serpentine hornbeam hedge.

Left: *Atlas* by sculptor Andrian Melka is framed by the hornbeam hedge.

Above: The Chinese Pavilion at the other end of the path.

On this occasion Stick happened to be carrying with him a piece of plywood to mark some planting plans, and he showed what he was thinking by taking a spray paint can and making a drawing on the plywood. In the end, the Moorish idea was abandoned, and the canal became a simple rectangular pool. In an unusual moment of practicality, they agreed that if Sleepy Cat Farm was ever sold, the next owners might prefer a lap pool rather than a Moorish canal with rows of water jets.

The seventy-five-foot-long pool thus became the centerpiece of the space along the west side of the Atlas garden, which is enclosed by a crenellated ilex hedge. A pebble mosaic terrace and a wall fountain, recently dramatized by a Bourdelle statue of a warrior seemingly floating on a bed of yews, punctuate the south end. A long exedra fronting a cat maze (a human maze downsized for cats) and a semicircular wisteria arbor create a magisterial focal point to the north.

Here, again, Charles Stick drew on his knowledge of garden-design history. "I am a great admirer of the Edwin Lutyens garden at Hestercombe in Somerset," he says. "I particularly loved a series of long rills with benches covered by trellised elms. I saw we could do something like that here." They could indeed, and it is one of the most successful elements of the garden, a place where people love to pause, sit in the shade, and take in the view. The wisteria arbor is huge, like the wingspan of an enormous bird. Its canopy protects the bench from the sun, and the bench, in turn, looks out over the low-growing boxwood maze. The whole composition is enthralling. "I've worked with wisteria for years," Stick observes, "and it is often very reluctant to flower. Fred's wisteria drips scented flowers all over the arbor in purple masses of blooms every spring."

Above: Charles Stick's on-site sketch for a rill, later changed to a reflecting/lap pool.

Right: In construction, the progression from an empty field to a painted circle for the cat maze, stakes for ilex border, and installation of ilex running to the distant wall.

Left: *Guerrier allongé au glaive* by Emile-Antoine Bourdelle, conceived in 1898 and cast at a later date, is installed on a block of stone surrounded by a low yew hedge.

Right above: The entry to the pebble terrace is marked by stone piers and a Chippendale-style gate.

Right: Potted Meyer lemon trees flank a chinoiserie bench.

Potted boxwood standards and ilex hedge complete the pebble terrace. The bronze lion head set into the panel above the fountain was designed by Charles Stick and cast by Kenneth Lynch and Sons.

Left and right below: Luxuriant wisteria, supported by a metal frame thirty-six feet wide and eleven feet deep, shades the curved bench behind the boxwood cat maze. At the center is a late eighteenth-century French carved-sandstone sundial.

Right above: Beyond the cat maze is the reflecting/lap pool.

This is the only part of Landman's landscape where two garden spaces act in parallel like separate "rooms," a design scheme much favored by English garden designers such as Vita Sackville-West at Sissinghurst in Kent, and by Americans like George Schoellkopf at Hollister House in Connecticut and Lawrence Johnston, who moved to England and made a garden at Hidcote in Gloucestershire that is one of the most visited in the UK. Stick would come to wonder about the decision to create these two side-by-side rooms. The repetition is, in a way, taking the control of space to an extreme. "I think about it," he concedes. "If two spaces are too similar, do people find it disharmonious?"

Above: Hidcote, in England, is a famous example of a garden designed in the form of rooms and connecting pathways.

Charles Stick once referred to the Golden Path Russell Page had created at Pepsico as "a necklace loaded with jewels." The expansion of the Landman's garden was coming together in the same way—linked spaces leading from one jeweled vista to another. With the completion of the Atlas and reflecting pool gardens, Fred Landman could only hope that an inviting slope that he was eyeing even further east would one day become his. Once again, he was fortunate, and in 2007 he was able to purchase a piece of property adjacent to the meeting between his property line and Lake Avenue. This acquisition moved the Sleepy Cat garden into an entirely new design direction.

Above: An aerial view reveals the visual connections and pathways that link the gardens at Sleepy Cat.

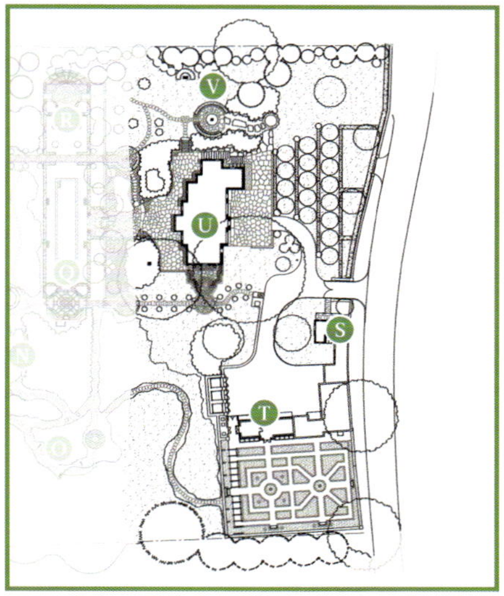

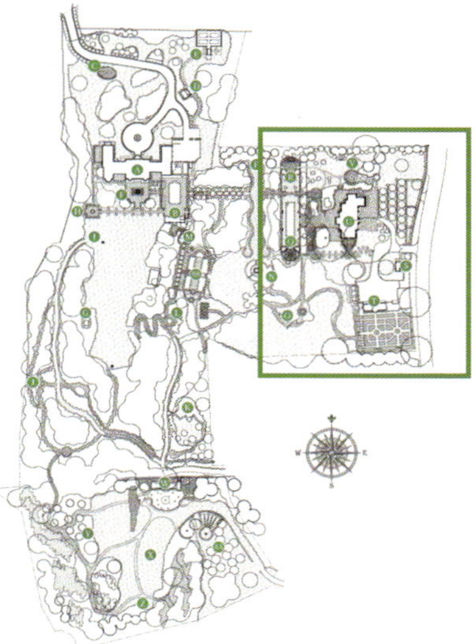

CHANGES IN STYLE

"I'm thinking vegetable garden; Fred's thinking parterre."

—SEEN LIPPERT

Fred Landman's marriage to Seen Lippert in 2003 changed many things about the garden. Lippert was not a gardener; she was a chef. She grew up in Fresno, California, and spent eleven years working at Chez Panisse with Alice Waters before moving with her two cats and two dogs to New York to start a restaurant and do some consulting. Shortly after they met, Fred Landman invited her to dinner in Greenwich and said he would cook. "Awesome!" Lippert thought. "Nobody ever cooks for me!" To her delight, the meal—a Marcella Hazan pasta recipe—was a great success. Landman, it turned out, had taken cooking classes in Italy.

He showed her the garden, of course, but it was neither his cooking nor his garden that sealed the deal. It was discovering that Landman lived with two cats and two dogs, just as she did. "That's how he stole my heart," she admits. The menagerie has grown considerably larger since then, becoming the inspiration for the name Sleepy Cat Farm.

Lippert saw her husband's garden as a perfectly lovely landscape, but how much of it was edible? She, after all, had grown up in the Central Valley, where gardens were farms, places that grew vegetables and fruits for people to eat, not landscaped with Golden Paths and parterres and pool pavilions and thousands of bulbs each spring.

Meanwhile, Landman and Stick were continuing to explore gardens abroad, now influenced by Lippert's own interests. Stick had shown the couple pictures of gardens he particularly liked, and Landman and Lippert went to see these, too. The upshot of these explorations and discussions always came back to the same issue—the need for more land. They had to find more space to fulfill their dreams.

Again, fate was on their side. In 2007 a two-acre parcel down a steep hill on the lower level of the eastern side of the property became available. This time the land came with several structures—a shingle-clad group dating from the 1920s including a main house, a one-room studio, and a service cottage, all adjacent to Landman's property line along the street.

This acquisition presented a different challenge to the Sleepy Cat team. Up until then, it had been a matter of corralling the ridge and woodland and turning them into landscapes that were often inspired by gardens seen in other countries or in books. The new acreage, with its existing structures, offered the opportunity to move away from the garden traditions the designers had so far adhered to and to create something very different. This was a project that fell perfectly into the wide-ranging portfolio of architect Chuck Hilton.

Preceding pages: The vegetable garden was conceived as a formal parterre, with boxwood edging the planting beds. Head gardener Alan Gorkin tends the beds.

"We found the buildings in very poor condition," Hilton recalls. "The combination of being built almost directly on the ground and suffering from decades of little to no maintenance left the roofs, siding, and fenestration needing replacement. The rooms were mostly a warren of little chopped-up spaces loosely organized around long-past utilitarian functions. Pretty much everything other than the foundation and super-structure needed repair or replacing." In spite of this negative assessment,

Above: Decomposed granite pathways are installed to divide the beds. One thousand boxwoods are set to define the pathways and beds.

Right: Charles Stick's plan for the vegetable beds and gourd arbor shows its intricate design.

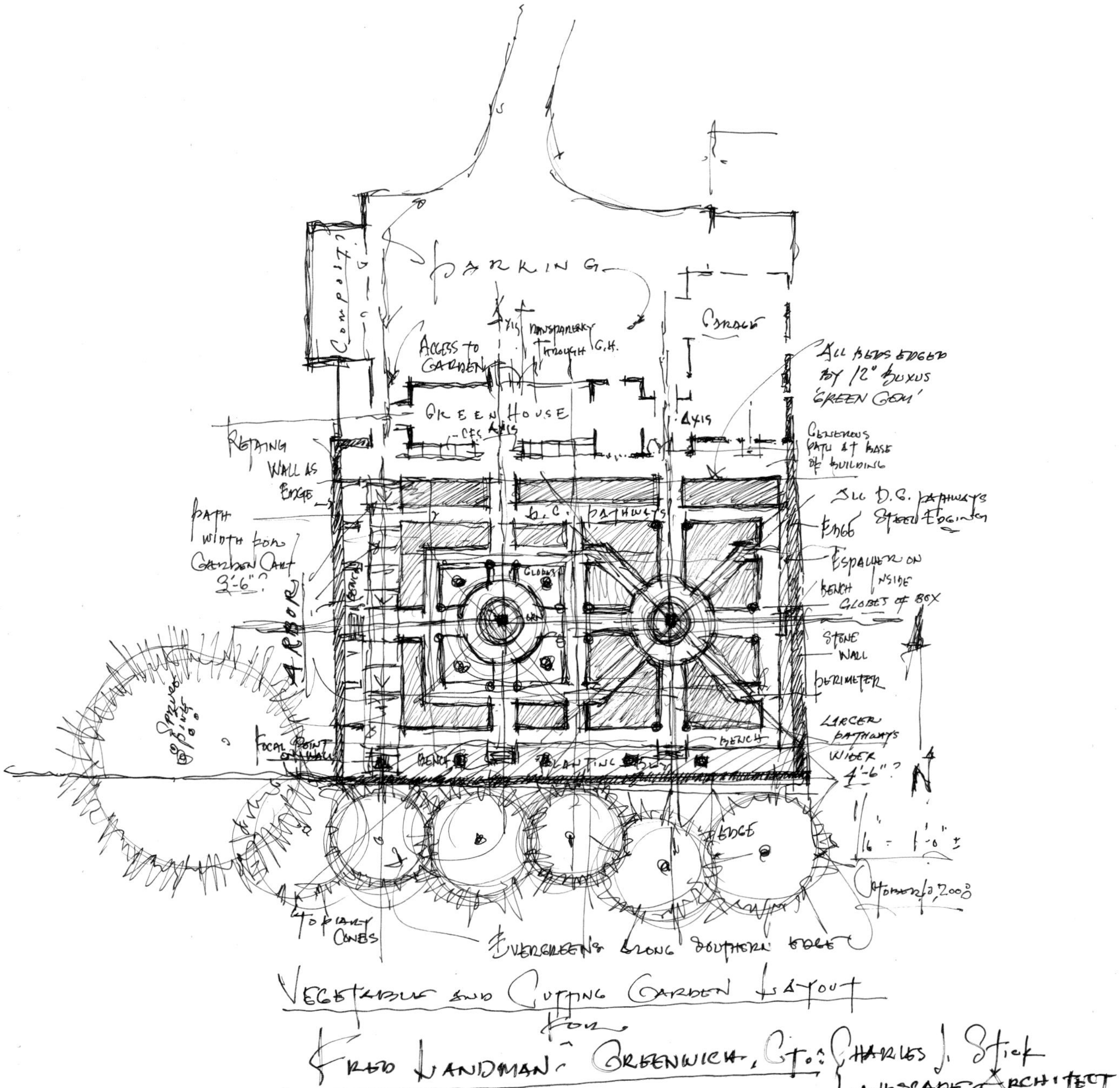

Hilton remained positive. "There was a certain quaintness and charm to the small awkward spaces that no one wanted to lose."

Debates about new plans for the garden extension now included the views of Seen Lippert. She saw the existing garden as a beautiful thing, of course. "I always found it very comfortable," she says. "But my idea of a garden was very basic. I was very naïve." She would see her husband and Charles Stick together in some part of the garden, heads together, talking intently, and then suddenly Stick would produce a paint stick and start spray-painting it. What on earth were they doing?

These mysterious exchanges had a great deal to do with what Hilton, Landman, and Stick finally decided to make of this space. Landman had already made a small vegetable garden, and Lippert thought this would simply be extended. She had no idea what was being planned. "I'm thinking vegetable garden; Fred's thinking parterre" is the way she put it. Chuck Hilton put it rather differently. "The idea was to create a small, organic working farm, which would augment the landscapes on the main property and produce fresh fruit and vegetables," he said. "It was a program that supported the owners' love of their gardens and their advocacy of fresh organic produce, and it also provided comfortable space for guests."

Hilton learned that the original designer was William Dominick, a Connecticut architect who was known to favor what he called the "French farmhouse style" for its "informality and picturesque qualities." Since Landman and his team had already agreed that the Georgian architecture of the main house was not suited to this collection of more rustic buildings, the idea of a French farmhouse vocabulary was very appealing. Hilton began to research French architecture that reflected the agrarian, vernacular style that the group was now dreaming about.

He studied many rural buildings in France, but perhaps the most significant was the hamlet Queen Marie Antoinette had built for herself in a corner of the vast grounds of Versailles. Smaller and more intimate than the Petit Trianon, this

Marie Antoinette's hamlet at Versailles displays the elaborate half-timber construction that inspired the Sleepy Cat farm buildings.

rustic retreat for the queen and her friends included a working farm with a dairy, a mill, and chicken coops. "For me, this small hamlet captured all the charm and character I was seeking, and after a few hundred pictures, I had all the reference material I could need," Hilton said.

Thus emerged the Sleepy Cat version of Marie Antoinette's retreat. With changes and much reduced, to be sure, but the atmosphere of the place is unmistakably

French and rural. The materials were often found locally with careful reference to French versions. The plaster came from France, the stained wood was distressed, the casework was antiqued (all by talented finisher Stephen Gamble). The addition of a cobblestone court, a greenhouse (imported from England), and a potting shed added to the overall feeling that one was stepping into a small corner of France. Chickens and bees were brought in to "people" this delightful new stage set and thus providing eggs and honey for Seen Lippert's many recipes. "Kitchen garden" as a practical description hardly does justice to what was finally created here.

Meanwhile, Charles Stick had not been idle. Farms, like any man-made landscape, require form and shape to grow successfully. Stick was involved from the beginning, designing a potager, an orchard, and a pattern of ornamental gardens framed in boxwood that grew cabbages and kale rather than poppies and petunias. Stick had designed a vegetable garden in Virginia, and he understood well the idea of the "ferme ornée," which had been promoted by Thomas Jefferson at Monticello. "The idea," he said," is to create something both beautiful and utilitarian." In this new garden, he was able to express both his love of formality and his respect for its most important function, producing fruit, vegetables, and herbs. His "workspace" was enlarged as his designs grew more daring, including an elegant arcade of cold frames and a gourd arbor that ran the entire length of the west side at the of the boxwood-lined network of patterned beds.

Fred Landman had also understood that the project was moving the garden into unexplored territory that was going to require more attention to planting programs and care than he could provide. He found his horticulturalist in Alan Gorkin, who had worked at the Brooklyn Botanic Garden and Wave Hill. He had the knowledge and background to supplement the Landman team. "Charles had a plot plan on paper," Alan says, "so I could take it home and fill in the blanks. I had free rein with the type of vegetables and seeds we worked with." Alan, with Peter Mauerman as his assistant, was later put in charge of all the horticultural aspects of Sleepy Cat Farm, including maintenance.

The collection of half-timbered farm buildings includes the chicken house with an octagonal chicken run and the potting shed between the garden office and greenhouse.

Left: The garden beds explode with a bounty of rare and familiar vegetables.

Right above: A greenhouse forms the north border of the walled garden with easy access to the vegetable garden.

Right below: *Hydrangea* 'Tardiva' blooms in Versailles planters.

Left: A mid-nineteenth-century armillary sphere mounted on a Gothic-style carved Bath stone column stands at an intersection in the vegetable garden. Jerusalem artichoke *helianthus tuberosus* is in bloom behind the bench.

Right: Artichoke plant *Cynara cardunculus.*

Far right: Bottle Gourd *Lagenaria siceraria.*

As for Seen Lippert, the whole concept was new to her. But now she could give Alan lists of herbs and plants she wanted, and there would be space for them. "I was missing things from home," she said. Rare vegetables and fruits that she regularly found in seed catalogs or brought back from travels abroad were given room to flourish. "Can this grow there?" she would ask, having found violets on the side of the road somewhere in the French countryside. Stick thrived on these challenges. Unlike in some parts of the garden, the land here was flat and easy to work with, and better still, he could climb up the hillside and look down at what he had wrought. His pleasure was evident when he took the measure of the vegetable gardens he had so carefully mapped out. "Even potagers can have structure," he observed.

Left above and below: Pear trees and grape vines are espaliered on the east wall above a stone bench.

Right: The arbor at the west edge of the vegetable garden is planted with edible gourds that envelope the frame in the late summer and fall.

Opposite: A head of Medusa and a limestone fountain are a dramatic focal point at the end of the gourd arbor.

Above: Flower of the *Cucurbita moschata*, *Zucchino* 'Rampicante', and various edible squash from Central and South America.

Vegetables grow in abundance
for chef Seen Lippert: cucumber
'Chicago Pickle,' tomato 'Celebrity,'
swiss chard 'Bright Lights,' beet
'Badger Flame,' onion 'Cabernet,'
pepper "Sweet Banana'.

Opposite: Rows of beehives line up in front of a cluster of rocks near the vegetable garden.

Above: A selection of the annual harvest from the garden, collected from bees, chickens and the good earth, are transformed with the help of Seen Lippert's expert touch.

The crowning glory of this part of Sleepy Cat Farm is the building that came to represent Fred Landman's salute to his wife. Known simply as the Barn, it replaced a small cottage on a slope not far from the farm cluster, and it was designed by Chuck Hilton with his usual attention to history and context. He took as his inspiration an illustrated book, *Manor Houses in Normandy*, which he and Landman studied carefully, searching for the perfect architectural style for the role it was to play. Norman architecture seemed to provide the warmth, rustic lines and careful detailing that would work best with the French vernacular farm complex.

Made of stucco and limestone, the Barn has antique half-timbered walls, a peaked slate roof, and Mauro Fidaleo's decorative exterior brickwork installed in a herringbone pattern familiar to travelers in Normandy. The interior is mostly a hymn to Seen Lippert's vocation, especially the ground floor, which is a 10,000-square-foot space with antique beams, heavy chandeliers hanging over a dining table for eighteen, opening into a huge kitchen with an island and an array of pots hanging from the exposed beams.

Opposite: Windows in the main hall of the Barn were inspired by those at Sir Edwin Lutyens's Deanery (1903).

Above: The kitchen and the root cellar.

Preceding pages: Rendering of the Barn and the completed building with its half-timbered facade and elaborate brickwork, tall windows, peaked slate roof, stone chimneys and imposing portico.

The form and details of the tower of the Barn are directly inspired by the pigeonnier at Les Quatre-Vents, Frank Cabot's garden in Quebec.

Another influence came into play here. Fred Landman had visited the house and gardens of Frank Cabot, a great gardener and the founder of The Garden Conservancy. Cabot's estate in Quebec, called Les Quatre-Vents, is in some ways similar in design to Sleepy Cat Farm, with its many themed gardens, pools and waterfalls. Landman instantly saw that Cabot's ideas could be applied brilliantly to his garden in Greenwich. In particular, he said, the pigeonnier in Cabot's garden inspired the tower of the Sleepy Cat Barn.

The gardens surrounding the Barn include another surprise. To the east side of the Barn, away from the elaborate quilts of herbs and vegetables, Stick laid out a small orchard—neat rows of 'Sugar Time' crabapples in strict geometrical form, delineated by boxwood hedges. Inspired by the ascetic designs of the Belgian landscape architect Jacques Wirtz, this exercise in abstraction is one of Stick's favorite spots.

Above: Rows of 'Sugar Time' crabapple interspersed with 'Green Velvet' boxwood hedges.

Opposite: A cascade beside the tower allows the sound of running water into the bedrooms. The hillside is covered with creeping raspberry *Rubus calycinoides*.

Aristide Maillol's *Ile de France* stands in the upper basin.

The staircase that leads to the Ting was extended by adding a slice down the hillside to the entrance of the Barn, where a huge statue of a River God stands guard. Stick worked closely with Chuck Hilton on all the design work, making sure, for example, that the placement of a big window at the rear of the Barn encompasses the River God outside.

Above: Rendering of the Barn in winter alludes to both the Deanery by Sir Edwin Lutyens and the cascade and River God at Ballyfin in County Laois, Ireland.

Opposite: The River God at Sleepy Cat faces a profusion of cherry trees (*Prunus yedoensis*) in full bloom.

One of the most spectacular visual contributions to the Barn complex, however, proved in the end to be quite simple. A thrilling sight in late spring and summer is the display of pink Eden roses climbing exuberantly up the exterior walls. Charles Stick had seen them growing at Orsan in central France (where they are called Pierre de Ronsard) and had never forgotten them. "They look lovely," he concedes. "But their loveliness is enhanced by their clinging to a wonderful French wall copied from a Normandy barn. Together they make an unforgettable image."

The development of these new areas of the garden have proved more than just a "beautiful and utilitarian" experience in the words of Charles Stick. With the additions of the extensive vegetable gardens and working greenhouses, Sleepy Cat Farm has truly honored its name and become a source of fresh food for staff and local families throughout the year.

Above left: Garlands of the Eden rose at Orsan, a priory garden in France.

Above right: The same rose looks just as happy growing up the walls of the Barn.

Opposite: The River God, resting on a stone plinth, leans on a jar that serves as a fountain spout.

Left: Aristide Maillol's *Pomona drapée* stands amid hostas, azaleas, and Japanese maple trees near the terrace beside the Barn.

Above: A nineteenth-century armillary sphere on a Portland stone sundial pedestal is set at the center of a bed of 'Tide Hill' boxwood.

Printemps sans bras and *Pomona drapée* are two of the five figures by Aristide Maillol that animate the Sleepy Cat landscape.

Opposite: Collections of boxwood in different shapes and sizes conjure images of a dragon with a spiked tail.

Above: A limestone obelisk is inscribed with an arrow pointing east and west, signifying the juxtaposition of garden influences.

Above: Steps lead to clouds of *Hydrangea macrophylla* 'All Summer Beauty.'

Right: Portrait heads, from left to right: Charles Stick, Fred Landman, and Charles Hilton.

Left: Two potted bonsai Japanese maple trees flank steps leading to the Japanese maple glade.

Right: A Chinese spirit on a rustic stone column is draped in the delicate leaves of Japanese maple *Acer shirasawanum* 'Moonrise'.

Left: Japanese maples display their full color: *Acer palmatum* 'Osakazuki' and bonsai *Acer palmatum* 'Kuro Hime'.

Opposite: The entrance to the perennial garden, with *Daphne genkwa* ablaze with lavender flowers.

Opposite: A dove tree (*Davidia involucrata*) and *Rhododendron* 'Pleasant White' contrast with the spring greens. The tree is often called a "handkerchief tree" because of the shape of its flowers.

Right: A life-size French bronze lion crowns the hillside at the top of the lion staircase. Flowers include narcissus, vinca minor, and *Helleborus orientalis* framed by magnolia and cherry trees in bloom.

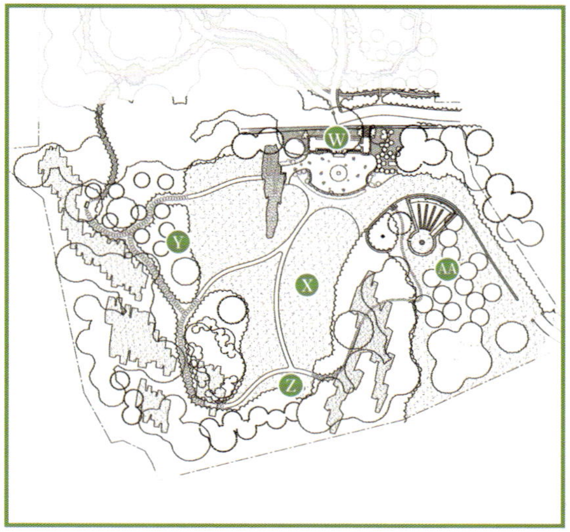

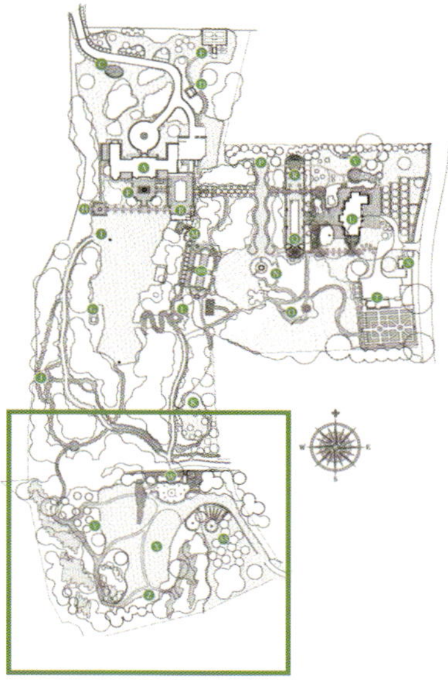

NEW LAND, NEW IDEAS

"A day in the garden" "A day in paradise"

—INSCRIPTIONS ON THE LIMONAIA COLUMNS

By mid-2008, the gardens at Sleepy Cat Farm had become far more than a series of beautiful spaces designed to enhance the main house. The built landscape was as important to the overall experience of the garden as the many glades and waterworks along the Golden Path, the elegant parterres and pool terraces, the Ting, the various flower gardens, and other horticultural delights now firmly in place. "Fred's a builder at heart," Chuck Hilton once said.

But to build, one must have land, and once again Fred Landman was lucky. In 2012 an additional three-and-one-half acres became available at the south end of the property. It was uneven, ungroomed land, with trees and rocks jutting out here and there, steep hillsides, and, dominating the north end, an abandoned storehouse. "Of course," said Landman, looking at the eyesore. "This is our Limonaia."

A limonaia is an Italian-inspired structure, usually of considerable architectural elegance, where citrus plants are stored during the winter months. Fred Landman was struck by many he saw in Italy, pointing in particular to the grand one in the Boboli Gardens in Florence. "It seemed to me to make sense to have a Limonaia," he says.

Chuck Hilton couldn't have agreed more. Once again, the architect turned to his reference books and wide travel experience. "We were limited to the existing footprint," he explains, "so the plan is long and linear with lots of openings on the south side. My inspiration for the facade was Palladio's Villa Barbaro, which I visited and admired while studying in Europe."

From the north side, the Limonaia looks slightly forbidding, even haunted. Built of masonry, it is reached by a bridge across a stream, shaded by a canopy of trees. Inside the mood changes. The interior is cool, a long rectangular space designed to house terra-cotta pots of citrus trees and other tender plants. A large stone sink, storage for garden tools, and a few chairs are the main furnishings, offering an opportunity for contemplation in a retreat from the stimulating vistas of the Golden Path.

Above: Invasive trees and an unsightly storage shed were cleared to make room for the Limonaia.

Preceding pages: The Limonaia seen from the meadow.

Right: Rendering of the Limonaia.

Below: Elevations of the Limonaia.

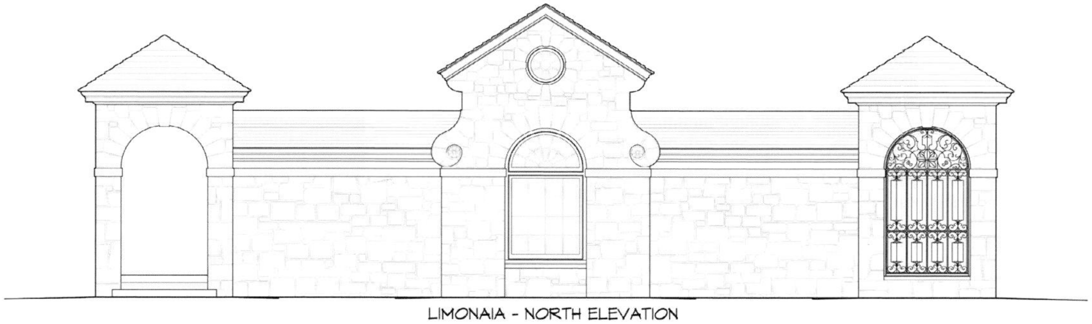

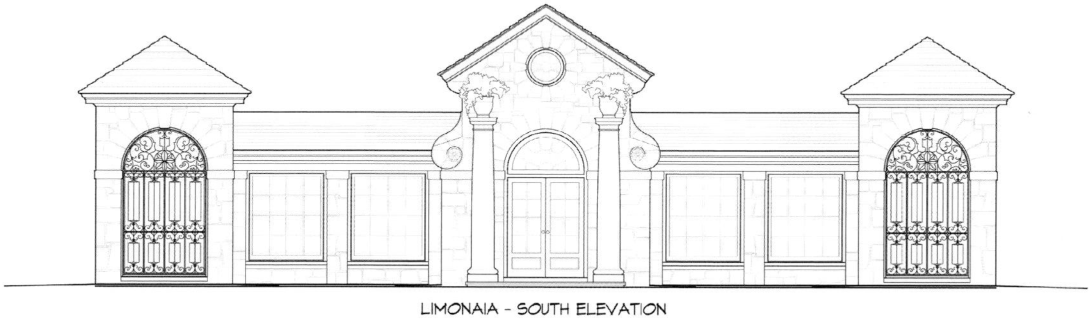

Above: Rendering of the urns that cap the columns flanking the entrance to the Limonaia.

Right: Eden roses climb luxuriantly up the walls.

Opposite: Lemon trees, with other fruit trees and agapanthus plants, are moved inside for winter storage. Bronze grillwork incorporates lemons as part of the Limonaia decorative motif.

The exterior is a masterpiece of lyrical sculpture and stonework, mostly thanks to the skills of master-mason Mauro Fidaleo. Other architectural details please the eye, including the elaborate bronze gates and decorative screens created by a local artisan. (Look closely and you will see lemons neatly entwined in the delicate bronze spirals.) But nothing is more pleasing than the terrace in the summertime, when the lemon trees are brought out of their winter quarters and bask in the sun alongside garlands of Eden roses, recently joined by Paul's Himalayan Musk, looping over the Palladian windows in a riot of fragrantly scented pink and white blooms.

The lemons, incidentally, are no ordinary lemons. They are Meyer lemons, considered the gold standard for citrus aficionados and challenging to grow in Connecticut. Seen Lippert had urged her husband to acquire some, and he and Stick found a perfect place to display them in the summer, rising gracefully from their terra-cotta pots

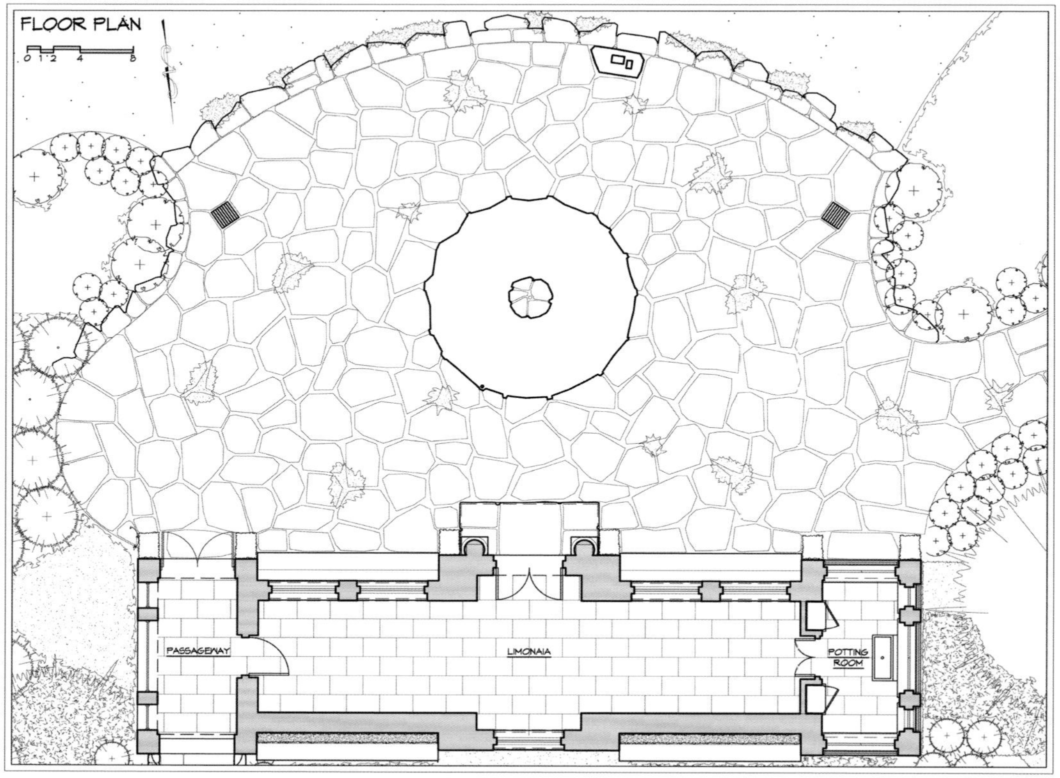

Above: Plan of the terrace.

beside the reflecting pool. But in the winter? Then the Limonaia comes into its own. Flooded with sunlight from the south, the masonry walls retain warmth, with special grow lamps installed to promote flowering of the Meyer lemons.

The area in front of the Limonaia was originally a mixture of meadow and natural outcroppings of rock (remnants of the ridge). In this sunny spot Landman and Stick decided to create a terrace, using slabs of natural cleft stone that would link it to the rocky outcroppings. Stick did the drawings as usual, but once the stone slabs were laid down, it was clear to all that the terrace was too big. "There should be a fountain there," said Landman. The two men looked at each other, facing the daunting thought of digging up the unwieldy paving stones, installing a pond and fountain, and then replacing the pavers. It was perhaps the only time a mismeasurement had caused such trouble, and Landman still teases Stick about it.

Above: The Limonaia in summer with Meyer lemon trees along the facade.

Right: Native waterlilies float in the circular basin at the center of the terrace.

Today, there is an asymmetrical stone terrace with agapanthus and other colorful plants mixed in the background and a pond and fountain as its centerpiece. One side of the terrace is bordered by a line of low-slung boulders, relics of the ever-present ridge. On the other side, wave upon wave of boxwood creates an enclosure between the terrace and the rambling meadow beyond. Rising up out of this green tide is the gleaming white statue of a woman bathing. "We put her in a sea of boxwood," Landman says.

A path leading from the terrace toward the meadow arrives at what is called the "Berry Wedge," a fenced-in plot (to protect it from birds) designated for the cultivation of all sorts of berries. To the right, slabs of walkable rock climb a steep hillside where beehives have been placed beneath fruit trees—peach, apricot, plum, apple. Honey, fruit, berries: a treasure trove of ingredients for Lippert's jams, jellies, and desserts.

Left: The terrace border is planted with *Phlox subulata* and *Euphorbia myrsinites*.

Above: Venus, modestly posed and surrounded by boxwood, gazes across the meadow.

The Limonaia creates a link between the French village buildings and the new territory opening up at the southern end of the property: the "Back 40," as they call it. This is another example of Landman and Stick's talent for incorporating the natural formation of the land into the overall design of the garden. The rocky ground, the wild meadow, the rolling hillsides, all played a role in creating a site where the romantic Limonaia would look its best, while offering the visitor to Sleepy Cat Farm yet another garden experience.

Above left: A Bacchante by Alice Cooper dances at the woodland edge of bald cypress, ostrich ferns, and camassia.

Above right: A garden seat inspired by William Kent's eighteenth-century seat at Rousham.

Opposite: A shallow stone stairway flanked by a massing of meadow plants— summer snow-flakes, Virginia bluebells, and hardy geraniums.

Above: The entrance to the Berry Wedge is marked by a Chinese quince.

Above right: Blackberries, raspberries, and strawberries flourish in the Berry Wedge.

Opposite: The Berry Wedge is fenced in with a "fedge," woven branches of the shrub willow *Salix miyabeana.*

Opposite: The trees in the orchard announce the summer: apples, pears, peaches, apricots, and plums.

Above: Senior gardener Peter Mauerman harvests peaches in the orchard.

Above: On the path to the Sacred Grove, a bronze figure of Echo, a replica of an Oscar Lenz original, stands against a background of massive rocks.

Opposite: This bench, nestled in the Sacred Grove, is a favorite place to sit and listen to the birds.

A group of five Ketton stone keystones, c. 1860, mounted on bronze poles with a boxwood base, mark a woodland path.

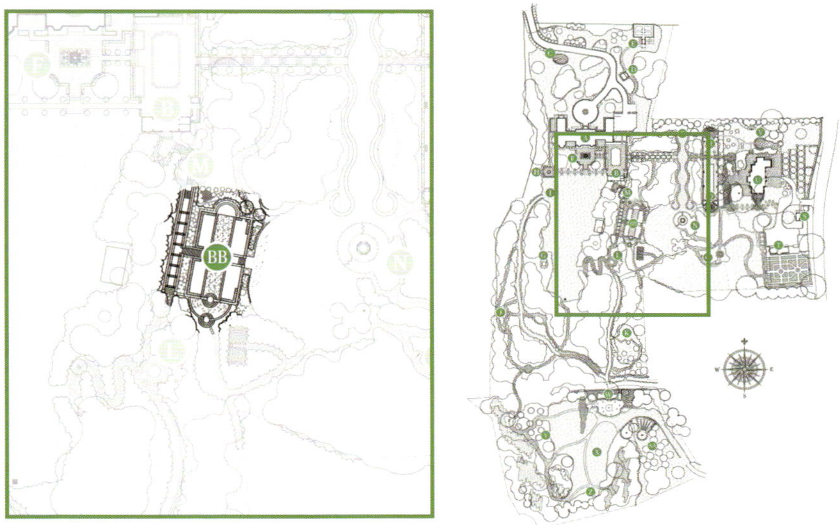

THE NEVER-ENDING STORY

"All gardening is landscape painting."

—WILLIAM KENT

The Limonaia was the last building to go up on the grounds of Sleepy Cat Farm. From then on, the spotlight would be turned full on to what is for many people the most important aspect of a garden—the flowers, plants, and shrubs that decorate the landscape. These are the elements that make up the great flower paintings we admire so much, from the Dutch still-life masters to Monet. And where is the most prominent display of flowers to be found in a traditional garden? The perennial border, of course.

The discussion of what makes a great garden continues to swing between the meticulous seventeenth-century French gardens—formal, symmetrical, tightly patterned—and the revolutionary English "landscape" garden, introduced a century later by William Kent in revolt against all that architectural rigidity. Kent and his contemporaries saw the landscape as a matter of rolling lawns, clumps of trees, lakes, and classical temples making up an idyllic, pastoral scene that Russell Page was later to scoff at. The Victorians also scoffed, deciding that the best gardening approach was to lay out flowers in strictly arranged beds, like quilts or coverlets (known as "bedding-out.") These garishly colored, linear carpets of flowers (mostly annuals) were all the rage in England by the end of the nineteenth century, and prizes were awarded to villages and towns all over the country for the "best" bedding-out of the year.

The perennial border was in part a response to this Victorian craze for micromanaging flowers. Perennial borders are by definition hardy; that is, the plants come back every year without having to be brought in from the cold. But they do not necessarily appear in the same place or in the same shape each year. In other words, they are very high maintenance and very badly behaved. This new, untamed, blowzy look became popular in England by the beginning of the twentieth century, promoted most enthusiastically by William Robinson and Gertrude Jekyll, brilliant gardeners whose ideas came to perfectly reflect this modern idea of plant design, focusing on a mixture of shrubs and flowers in loose drifts of color, large and small blooms, in all shapes and sizes, jammed together and spilling over untrimmed edges in oversized flower beds. This kind of planting style worked particularly well in the UK, with its moderate climate, incessant light rain, and reaction against too much order.

Preceding pages: The perennial border in the dappled late afternoon light.

After so much meticulous design planning with the allées, boxwood rows, steep staircases, and undulating hedges that dominate the Sleepy Cat garden, perhaps it's not surprising that Landman and his team came to dream about a perennial border. Time now for some uncontrolled floral beauty, they thought. Fred Landman had visited Great Dixter in Rye, where he was exposed to one of the most famous perennial borders in England. "We could make one of those," he said with his usual optimism.

Above left: The perennial long border at Great Dixter in England, an inspiration to gardeners all over the world.

Above right: The old orchard was replaced by a late addition, the perennial border.

"We had the space," Charles Stick agreed. "There was quite a steep slope on which a not particularly impressive small orchard of trees grew." So during the winter of 2016–17 the team embarked on building a perennial border. "The problem was levelling the ground," Stick recalls. "To do so, it meant building a nine-foot wall." Mauro Fidaleo again came to the rescue and, with his team, moved thirty tons of rock to make a wall that would contain the soil.

Of course, it was never going to be very like Great Dixter. Most of the great English perennial borders are extremely long and wide, the beds sometimes double-dug in width, stretching almost as far as the eye can see, filled with a torrent of flowers and shrubs in various form, shapes, and colors. "It was overwhelming," Fred Landman admits. "You feel they are going to overtake you."

The Sleepy Cat version was restricted by very limited space. "It's a long, linear feature that could be longer and wider, but the effect is still pleasing," Stick comments. One unexpected bonus of the placement of the border is its height, revealing splendid views of the transition down to the Golden Path and the lawn rolling out toward the Ting. An added feature was a "ruin," a stone arch at the south end of the border, reflecting the fascination of many English garden designers with artificial ruins.

Left: During construction of the perennial border planting beds, thirty tons of rock were removed and twenty tons of new stone and soil were brought in. Then plants and hedges were laid out for planting.

For Fred Landman, the development of the perennial border was a new experience. The questions he had to ask were very different from those he had been accustomed to. "What is going to bloom here?" When? Should it be bigger? Smaller? Does it like it here? Should we move it?" One doesn't have to ask that of boxwood. Landman can be an impatient gardener, and the slow growth of many of the perennials in his new border frustrated him. "Will it flower in my lifetime?" he would teasingly ask horticulturalist Alan Gorkin.

Left: Steps at Les Quatre Vents inspired by Sir Edwin Lutyens.

Right: Fred Landman, Charles Stick, and Mauro Fidaleo discuss the finishing touches to the perennial garden steps and hardscape.

But the waiting soon began to pay off, and then, as often happens with Landman, he became committed. Annuals, once a perfectly acceptable part of the garden planting program, were anathema. "Annuals are cheating!" he declared. "No annuals!" He has since modified his position. After all, Gertrude Jekyll had no shame in using annuals to fill holes or provide a burst of color in her borders.

But it is, of course, a work in progress. All the great perennial borders in the world are works in progress. Perennials die. They grow too big. They get lost. They have to be replaced. Their blooming time alters. The sun hits them earlier—or later. As the climate changes, winters are more extreme. And so on. "The goal," Alan Gorkin says, "is to have as long a period of bloom year-round as possible." That goal does not get achieved overnight, if at all. Fred Landman by now has learned that.

PORTA
PARADISO

Opposite: The steps and arch at the south entrance to the border.

Above: Rendering of the border and the arch.

Perennial Long
Border Garden

Opposite: The east staircase to the perennial border is planted with *Gladiolus byzantinus, Salvia* 'Indigo Spires', and *Miscanthus sinensis* in pots.

Above: *Dichondra* 'Silver Falls' and *Coleus* 'Wedding Dance' are some of the plants on the steps.

Left: An armillary sphere set on a nineteenth-century Italian carved-marble pedestal stands near the path leading to the prayer wheels.

Opposite: Asiatic lilies 'Pink Perfection' add color to a stone pillar above the perennial border.

Opposite: Lilies and delphiniums mingle with salvias in front of the perennial border wall.

Above left: Dahlias are always showy contributors to a perennial border.

Above: *Allium giganteum* and *Baptisia* create height and shape in the border.

The turtle pool at Heywood Gardens in Ireland, designed by Sir Edwin Lutyens, 1912.

Landman was exposed to many indelible experiences and occasions of reflection during his long garden journey, but none more poignant than in 2012 when he suffered a massive heart attack and subsequent heart transplant in Italy. During his year-long recovery, Landman would slowly stroll through Sleepy Cat gardens, always therapeutic and beautifully regenerative. It took over a year before he could walk properly again, and during that time he began to think about the future of the garden and how best to preserve it for future generations.

When he was ready to travel again, the place he most wanted to go to was Ireland. "I'd read about a small garden outside Dublin designed by Edwin Lutyens in 1912, and I wanted to see it." He went, and he was not disappointed. "It was a jewel of a garden," he remembers. "Not elaborate or grand. Part of an estate long gone. Just a little gem created over a hundred years ago. I fell in love with it."

The name of the place was Heywood Gardens, and unlike most of the gardens Landman had visited, this one had no building to relate to—the house had been razed many years earlier. But fifty acres of parkland remained, decorated with terraces, stone walls, obelisks, allées, and other architectural features, along with perhaps one of Lutyens's least-known masterpiece, a sunken garden. As Landman studied the garden's impeccable lines, its pared-down simplicity, its lightness of color and texture, he felt transformed. This little sunken garden in Ireland transported him to another level of understanding about himself and his project of a lifetime back home.

Lutyens made his garden more than one hundred years ago, and yet its vitality and elegance remained unquenched, its spiritual power still almost tangible. "Now I know what a master gardener can do," Landman said to himself, returning to his own garden with renewed enthusiasm. "Lutyens's genius—and it is genius—spurred me on. I now could think about my garden and its long-term future. I set up and funded the Sleepy Cat Farm Foundation and started the process of transferring the operation and assets of the garden into the Foundation as a means to ensure its future."

These days, walking through the garden as he has always done, Landman discovers new areas where he wishes to experiment and mulls over ideas recalled from all his garden visits over the years. "I am more demanding now," he says. "When it's supposed to shine, it must shine." Plans continue to crowd the desks of the Landman team. More than twenty statues now stand in the garden. The fifth Maillol is the latest to join the list. It has a possible site picked out, but with typical attention to detail, Landman wants to make sure it will be placed so that visitors can see it from all sides, "just as it should be seen," he says.

Sometimes Charles Stick comes up from Virginia, and the two walk around the garden together, observing, discussing, assessing, fine-tuning. As Landman says, "There is no end to making a garden. You will never reach perfection." In the past few years, he and his team have planted over 500,000 bulbs. They will still be planting next year, and the year after that.

Working with nature often encourages thoughts of the future. The impact of Lutyens's small garden in Ireland has lasted more than a century after his death. There is every reason to suppose Landman's own vision at Sleepy Cat Farm will one day be regarded as an equally inspiring contribution to posterity. His wife, Seen, has said that on their many travels to look at gardens, "Fred would plant seeds in his mind of what he had seen," and then when they got home, he'd water them—surely a recipe for some kind of immortality.

Left: A tantalizing glimpse of the Chinese Pavilion from the perennial border.

Opposite: The perennial border gate pays tribute to the *porte* in Italian city walls. Limestone panels above the arch on both sides are inscribed "Porta Paradiso"— Gate of Paradise.

PORTA
PARADISO

AFTERWORD

I wasn't always a gardener. Working full time with a growing family, I managed to be at most a weekend gardener. On Saturday mornings with spade and trowel in hand, I would visit the local garden center and pick out plants that caught my attention. Every spring a small vegetable garden, a few plants, shrubs, and the occasional tree would find their way into the ground.

While working, my mentor had a saying, "Reach for the stars, always think big." Over a twenty-seven-year period, I went from Junior Research Assistant to CEO, guiding a start-up global communication satellite company. Literally reaching for the stars, it eventually became the largest privately held global communication satellite company of its kind.

What I learned from this work experience is that if you can assemble a team of experts in their respective fields, you too can succeed. This group taught me how satellites worked, the regulatory environment for providing satellite services around the globe and market demand for video and data services, in each and every country. After an exciting sixteen years putting up more than fifteen satellites covering 98 percent of the world population, we transitioned from a private company to one listed on the NASDAQ.

Now, it was time for me to look for a new challenge. I had bought a big house in Greenwich that cried out for a garden. Starting from scratch, I decided to turn the existing ground into a series of landscapes, using all the ideas I had seen in my travels to create a modest paradise. That same "think big" mantra carried over to the garden venture. Like the satellite business, I assembled a great team. No longer was it satellite communication engineers, lawyers, and economists; it was horticulturists, architects, landscape architects, masons, and builders.

The conversation switched from satellite solar arrays, geostationary orbital positions, and launch vehicles to drainage plans, stone materials, garden "rooms," pathways, and plants. There was no grand master plan; one thing led to another, as vistas opened up and land became available. Working with this creative new team of experts transformed me from a weekend gardener to full-time, seven-days-a-week gardener.

Now I am a gardener both at home and away. Before, my travels were about counting satellite dishes and attending business meetings, with the occasional visit to a museum and hopefully a great restaurant. My new focus became visiting important gardens, both large and small, in Europe and elsewhere. I absorbed my surroundings and studied what made these gardens unique and inviting, what different styles and plantings were used. Seeds, especially heirloom, became a staple for me to bring home to plant new varieties in the vegetable garden. I always returned inspired by something I saw or experienced, and working with my new team, we set about integrating and adapting some of these ideas into Sleepy Cat Farm. Gradually the landscape evolved into a series of experiences that visitors today find both pleasurable and inspiring.

Sleepy Cat Farm represents more than two decades of travel and observation, many trials and a few errors, the creative efforts of a talented team and my personal expression of a truly unique garden. As it grew, I grew with it, and I suspect I will continue my journey of becoming a gardener as long as I am able. There is no more gratifying experience than to build, in cooperation with others, a place of beauty that will last for generations to come.

Fred Landman, January 2021

CONTRIBUTORS

Fred Landman

The beneficiary of the talented people on these pages, who allowed me to fulfill my dream of a beautiful life in a beautiful garden.

Seen Lippert

My wife, who has an appreciation of the garden but a deep passion for what ends up in the kitchen. Whether it's peaches for cobbler, rose petals for syrup, or beautiful lettuce for a simple salad . . . all is put to good use.

Charles Stick

I've been a student of Charles's course "Making a Garden" for more than twenty years, and I could not have asked for a better teacher and mentor. A true classicist, he not only knows what works best, but he can tell you why.

Charles Hilton

Precedent is critical to what and how Chuck designs projects. We've collaborated on a Georgian brick house, a French Normandy barn, our Italian auxiliary building, as well as a prewar New York City apartment. In all cases, Chuck did the research and set the appropriate direction.

Isabelle Vanneck

Our kinship began twenty-five years ago with a single piece of furniture. Since then, we have collaborated on all the residence sites in sync. She has brought an aesthetic that I admire and appreciate.

Alan Gorkin

Alan has an amazing knowledge of plants, fruit trees, propagation methods and how to make vegetable gardens. He arrived at Sleepy Cat Farm in 2009 and quickly brought our horticulture to a new level.

Mauro Fideleo

Mauro is truly a master mason with deep knowledge of traditional stonework. Each brick, stone, and masonry structure reveals his artistic touch.

Larry Labriola

Larry's company, Carmine Labriola Contracting Corporation, has provided the resources to do major planting installations as well as lawn and hedge maintenance for twenty-two years. He continues to keep the property looking its best.

Robert Levine

"Bob the builder," as I called him, assembled, and supervised the team of people who made these award-winning buildings. A best friend, and the best man at my wedding, he is sorely missed by many.

GARDENS TO VISIT

Chatsworth
Bakewell, Derbyshire
DE45 IPP
United Kingdom
+44 1246 565300
www.chatsworth.org/gardens

Cliveden
Taplow Road
Maidenhead SC1 8NS
United Kingdom
+44 1628 668561
www.nationaltrust.org.uk/cliveden

Giardini La Mortella
Via Francesco Calise 45
80075 Foria d'Ischia (NA)
Italy
+39 081 986 220
www.lamortella.org

Great Dixter
Great Dixter Driver
Northiam, Rye, East Sussex TN31 6PH
United Kingdom
+44 17978 252878
www.greatdixter.co.uk

Hestercombe
Taunton, Somerset
Cheddon Fitzpaine TA2 8L6
United Kingdom
+44 1823 413923
www.hestercombe.com

Heywood Gardens
Ballinakill Road
Portlaoise, County Laois R32 K4V6
Ireland
+353 57 873 3563
www.heritageireland.ie

Hidcote
Hidcote Bartrim
Chipping Campden
Gloucestershire GL55 6LR
United Kingdom
+44 1386 438333
www.nationaltrust.org.uk/hidcote

Hollister House Garden
300 Nettleton Hollow Road
Washington, Connecticut 06793
860-868-2200
www.hollisterhousegarden.org

Les Jardins de Quatre-Vents
135 Boulevard Malcolm-Fraser
La Malbaie, QC. 65A 1A2
Canada
+1 418-434-2209

Moss Temple (Kokedera)
5 6 Matsuojingatanichou
Kyoutoshinishikyou-ku,
Kyoto 615-8286
Japan
+81 75-391-3631

Orsan
Prieure d'Orsan
18170 Maissonnais
France
02 48 56 27 50
www.prieuredorsan.com

Pepsico
700 Anderson Hill Road
Purchase, New York 10577
914-253-3150
www.pepsico.com/sculpture-gardens

Rousham
Station Road
Rousham OX25 4QU
United Kingdom
+44 1869 347110
www.rousham.org

Sissinghurst
Biddenden Road
Cranbrook, Kent TN17 2AB
United Kingdom
+44 1580 710701
www.nationaltrust.org.uk/sissinghurst

Valsanzibio
Via Diana 2
Valsanzibio di Galzignana Terme
35030 Padua
Italy
+39 049 913 1065
www.valsanzibiogiardino.com

Vaux-le-Vicomte
Chateau de Vaux-le-Vicomte
77950 Maincy
France
www.vaux-le-vicomte.com

Villa Barbaro
Via Comuda 7
31010 Maser (Treviso)
Italy
+39 0423 923 004
www.villadimaser.it

Villa Carlotta
Via Regina Teodolinda 2
Tremezzo, Lombardy 22016
Italy
+ 39 0344 40405

Villa Chigi Cetinale
Strada di Centinale 9
53019 Souicille (Siena)
Italy
+39 0577 311147

Villa La Foce
Strada della Vittoria, 61
53042 Chianciano Terme
(Siena) Italy
+39 0578 69101
www.lafoce.com

Les Jardins de Quatre-Vents

Photographs by Curtice Taylor
except as listed below:
Jon Connor 78 left
Gus Ford 60-61
Charles Hilton Architects 20, 22 left and right, 23, 36 left, 107, 109, 111 top, 123, 124, 125 left, 127 right, 148, 150 right, 188 right
Fred Landman 11 left, 100 right, 105 left top and left bottom, 130 left bottom, 120, 172
Kathy Landman 12, 13, 36 right, 44, 70 right, 81, 83, 101, 113 right, 125 right, 150 left, 156, 161, 166-67, 171 right, 173 right, 186, 188 left, center left, center right, 189, back cover, Curtice Taylor portrait
Karen McLean Caroline Seebohm portrait
Charles Stick 42, 52, 53, 56 left, 68 top and bottom, 70 left and center, 86 right top and bottom, 88, 92, 93, 100 left, 127 left, 130 top left, 132 left, 171 left, 182, 191
Aerial photography by Stefen Turner (all rights reserved) 14

Illustrations/Drafting:
Patricia Curtan 24 right
Paul Gaertner 6
Charles Hilton Architects 17, 37 bottom, 45, 102, 122, 151 center and bottom, 154
Peter Lorenzoni 56 right, 64 top, 69, 130 right, 151 top, 153 left, 175
Karen Meyerhoff Endpapers
Office of Charles Stick 35, 45

First published in the United States by The Monacelli Press, a division of Phaidon Press.

Library of Congress Control Number 2021910924
ISBN 9781580935814

Design: Susan Evans, Design per se

Printed in China

The Monacelli Press
65 Bleecker Street
New York, New York 10012

Mini
Missy
Matta
Macy
Marcella
Moko and Matta
Marlon
Midas
Mitsou
Max
Lola
Maillol